THE NATIONAL POLICY AND THE
WHEAT ECONOMY

→»«←

SOCIAL CREDIT IN ALBERTA

Its Background and Development

A series of studies sponsored by the Canadian Social Science Research Council, directed and edited by S. D. Clark. The series is now complete.

THE NATIONAL POLICY
AND THE
WHEAT ECONOMY

Vernon C. Fowke

University of Toronto Press

TO MY MOTHER

for her vision and courage

->»«<-

Foreword

WITH THE PUBLICATION of what is now the seventh of a series of ten studies sponsored by the Canadian Social Science Research Council relating to the background and development of the Social Credit movement in Alberta it might appear becoming to apologize for the delay in bringing to completion a research project launched as long ago as 1943. But no apologies are offered. Scholarly books like this one of Professor Fowke's are not something which can be produced overnight. Much hard and patient work and thought are required, and much time. The publication of the present study offers thus the occasion not for apology but for an expression of appreciation of the patience and understanding of the project's sponsor. The increasing dependence of the scholar upon financial assistance in the carrying on of research has brought with it many dangers, not the least of which is the working to a dead-line. Too often organizations granting money for research expect —and, indeed, demand—results in a certain specified time. What the scholar produces becomes in effect a commodity to be bought, delivery to be made on a certain agreed date.

Given the fact that research in the social sciences has become an expensive business, it is perhaps inevitable that much of it will be carried on under conditions determined by "good accounting practice." The question might well be raised, however, whether we are as dependent upon outside financial assistance as we have made out. The argument that such assistance is necessary if the big problems in the social sciences are to be studied has merit, but we need to ask whether it really works out this way. With certain important exceptions, the effect of outside financial assistance has been to steer scholars away from the study of the big to the study of the small problems of our science, to the study, in other words, of those problems simple enough to make possible the collaboration of a number of people, and of such a limited scope that they can be investigated in a short period of time. Few scholars today have the courage to embark upon a piece of work which is going to take them ten or fifteen years—or a lifetime—to complete. *The Fur Trade in Canada* could never have been written if progress reports had been expected every six months and the finished work in two years! Professor Fowke, and the authors of the other studies in this series, might be embarrassed if it was reported how very few dollars went into their studies. Not embarrassing, however, would be an indication of how much hard work and thought over the years were necessary before these studies could be completed.

Research of a kind can be bought but scholarly work cannot, and no organization has been more aware of this than the Canadian Social Science Research Council. With government and business becoming more interested in the support of research in the social sciences, however, there is good reason for concern regarding the conditions under which scholarly research may be carried on. It would be presumptuous to set forward the Social Credit project as a model to be followed, but something may be learned by the experience in carrying out this project. Certainly, this experience offers a convincing demonstration of how much can be accomplished by a small expenditure of funds if care is taken to preserve those conditions which make for scholarly work. It would be invidious to single out for mention in this regard any one of the studies which has formed a part of this project. The authors of the other studies would certainly forgive, however, a word here in conclusion giving expression to the view that few published works more fully demonstrate the values of true scholarly research than this study by Professor Fowke.

S. D. CLARK

Contents

PART ONE

ORIGINS AND DEVELOPMENT OF THE NATIONAL POLICY

CHAPTER ONE

The National Policy: A Matter of Definition

CANADIANS are familiar with the term "National Policy." It refers specifically to the policy of tariff protection which was put into effect by the Conservatives in 1879 after their victory in the federal elections of the preceding year. Dr. Skelton has indicated the evolutionary nature of the concept. He describes the expression, "National Policy," as "the phrase which Rose devised, Hincks stamped with his approval, and Macdonald made current."[1] He points out that "early in Confederation there were repeated tentative gropings towards the National Policy. . . ."[2] His interpretation is that the National Policy of protective tariffs emerged early in the Confederation period because it was essential to the accomplishment of the economic and political objectives for which Confederation was established. Macdonald proposed the adoption of the National Policy in a resolution presented to the House of Commons in 1878 while the Conservatives were in opposition.[3] The policy was a major plank in the platform on which his party won the election of 1878.

This particular meaning of the term, National Policy, with its origins in historical accident, has become firmly rooted in Canadian thought and expression. For Canadians the National Policy means Macdonald's tariff policy. The capitalized phrase will undoubtedly long continue to designate the historic milestone at which Canadians abandoned the idea of tariffs for revenue only, discarded even the euphemism, "incidental protection," and deliberately set foot on the pathway marked "protection."

Firmly rooted though it be, the usage is unfortunate. National policy ordinarily means simply the policy of a national government. As such,

[1] See O. D. Skelton, "General Economic History of the Dominion 1867–1912" in A. Shortt and A. G. Doughty, eds., *Canada and Its Provinces* (Toronto, 1914–17), vol. IX, p. 146.

[2] *Ibid.*

[3] On March 7, 1878, as the House moved to go into committee on the budget, Macdonald introduced the resolution that "This House is of the opinion that the welfare of Canada requires the adoption of a National Policy, which, by a judicious readjustment of the Tariff, will benefit and foster the agricultural, the mining, the manufacturing and other interests of the Dominion." See Canada, *House of Commons Debates*, 1878, vol. I, p. 854.

it is as comprehensive as the field of constitutional competence of the government in question. It is specific to specific matters, but what these are in a particular circumstance only the context of discussion can indicate. Thus, for example, we should be able to speak of the national policy regarding defence, transportation, control of trade, economic expansion, or any one of a hundred other items. No nation would tolerate a national government which for generations had but a single policy, the imposition and maintenance of protective tariffs. Canadians know well that tariff protection has been but one of a countless number of the policies of their national government.

Governmental policies, however, vary markedly in relative importance and though the variations may not be precisely measurable at any point of time, certain broad differences can readily be noted. This is particularly true when the policies in question are grouped according to their combined major purpose. Some policies will clearly relate to minutiae or to temporary contingencies only, while others will be integrated functionally toward the furtherance of fundamental and persistent governmental aims. The decision taken by the Canadian government to foster the industrialization of the country by means of tariff protection was one of the latter type of policies. It was of considerable importance in itself, and, along with the group of policies of which it formed a functional part, it was directed toward the accomplishment of the most basic and persistent of the purposes of the Canadian government. In this regard it must be emphasized that the national policy predated the creation of a national government in Canada and envisaged the establishment of such a government as one of its indispensable instruments.

The central purpose of the Dominion government, the task for which it was created, was to form and develop in British North America a political and economic unit on a national basis. The areas out of which the national unit was to be formed comprised some or all of the diverse and geographically dispersed British provinces, colonies, and territories in North America. The basic intention was to unite these areas politically and economically in order that they might more effectively cope with alterations which had taken place in imperial and continental relationships and which rendered existing arrangements obsolete if not wholly inadequate.[4]

[4]Hon. Thomas D'Arcy McGee made it clear that some at least of the Fathers of Confederation regarded the purposes of Confederation as diverse and not exclusively political or economic. See quotation, chap. III, pp. 33–4.

Political alterations which pointed urgently to the necessity for change in the relationships among British North American territories appeared both internally and externally after 1850. Internally there was the political impasse in the United Province of Canada arising out of rigidities created by the Act of Union. These rigidities, originally immaterial, became evident when unexpected shifts occurred in the regional and racial distribution of population within the united area. Externally, there were new and disturbing developments in the matter of defence which applied alike to the St. Lawrence and to the maritime provinces. After 1850 the Imperial government displayed an increasing reluctance to maintain the burden of defence of British North American territories. This development coincided with a period of increasing friction between Britain and the British North American provinces on the one hand and the United States on the other. The central government, to be created by the act of Confederation would, so it was reasoned, assume responsibility for defence on a national basis, otherwise the defence requirement would either impose an intolerable burden on the individual provinces or else would have to be left in default. Federation would likewise provide an escape from the political impasse by permitting the principle of representation by population to be applied in the field of national legislative action while leaving the furtherance of purely local purposes to the care of provincial governments.

The economic forces which contributed to Confederation and the establishment of the Dominion government were, like the political forces, both internal and external. The decision to create and develop an integrated economy on a national basis was adopted because of the disappearance of not one but two more highly regarded alternative possibilities—those of imperial and of continental economic integration. The alteration of British trade and fiscal policy at mid-century as represented by the removal of imperial preferences not only ended any hope of the immediate extension of imperial integration but also seriously reduced that which had previously existed. The Reciprocity Treaty, put into effect in the 1850's with the United States, restored temporarily to full vigour the persistent belief among British North American economic groups that continental integration offered a practicable and preferable alternative to close imperial economic relationships. This belief, however, waned in proportion as the conviction grew that the Americans were unlikely to renew the treaty upon the expiration of its first term.

In turning from continental to national economic objectives the

Anglo-Canadians on the St. Lawrence abandoned—or appeared to abandon—a quest of long duration.[5] From the days of its first European occupation the St. Lawrence had been regarded as a continental trade route, a means of access to the trade and commerce of the interior. The French fur traders had fully utilized its potentialities and after the fall of Quebec the Anglo-Americans, with headquarters in Montreal, had extended the fur-trading empire to its limits on the Arctic and the Pacific. This unity ended with the absorption of the North West Company by the Hudson's Bay Company in 1821. American national expansion also contributed to the decline of the North West Company by an increasing curtailment of that Company's southern and southwestern fields of operation.

Agricultural expansion westward from the Atlantic coastal regions was the most important element in North American economic life in the decades after the Revolutionary War. The expansion was continental, and the settlers, whether from the older settlements to the eastward or from across the Atlantic, paid scant attention to the invisible political boundary which was projected across the continent in step with American national aspirations. Agricultural expansion meant the expansion of trade and commerce in agricultural products and in agricultural supply. Potash and wheat were among the first products to move commercially from the agricultural frontier. Between 1790 and 1850 new agricultural settlement enveloped the lower Great Lakes in its westward advance and provided North America with its first wheat economy and with its first staple-producing agricultural community. The commercial counterpart of agricultural export is agricultural supply.

Regardless of the distinction between agricultural export and agricultural supply, two points must be emphasized here. First, the agricultural commerce which succeeded the fur trade in northeastern North America over the turn of the eighteenth and nineteenth centuries was potentially continental rather than national in scope. Second, the mer-

---

[5]Confederation did not end the attempts on the part of Canadians to secure a restoration of reciprocity with the United States. The attempts which were made were not necessarily directed toward the establishment of complete free trade between the two countries. The Reciprocity Treaty of 1854 had provided for reciprocal free trade in natural products only and had made no provision for adjustment of tariff rates on manufactured goods. The Reciprocity Treaty which Canadians rejected at the polls in 1911 was not intended to secure complete free trade between the United States and Canada. Various official Canadian "pilgrimages" sought a reciprocity agreement in Washington between 1854 and 1911, but reciprocity was not regarded as synonymous with free trade. Only certain proposals put forward in 1887 and 1888 envisaged complete free trade. See O. D. Skelton, chart entitled "Commericial Negotiations with the United States, 1854–1911" in *Canada and Its Provinces*, vol. IX, facing p. 126.

chants in Canada regarded the St. Lawrence as the "natural" trade route for the agricultural commerce of the interior as they had for the continental fur trade. At the very least, so they reasoned, they should be able to share in the agricultural trade of the American frontier regardless of political boundaries. Such reasoning was challenged—and effectively challenged—by American merchants. Public and private developmental expenditures improved the American trade route—the Hudson–Mohawk river system—first by the construction of the Erie Barge Canal and later by the construction of a system of railways which joined the American Middle West with New York. Efforts to make good the Canadian claims in this matter fostered the construction of St. Lawrence canals and the Grand Trunk Railway, burdened provincial governments with heavy debts,[6] and created serious constitutional difficulties. Since the goal sought was competitive competence in relation to American interests, the persistent failure to establish that competence left only the difficult choice between accepting defeat and instituting some boldly imaginative new method of attack upon the problem.

The English colonies which created the American national state in the late eighteenth century did so after they had willingly withdrawn from the imperial economic framework. The national government was assigned the duties of defence and economic development which the colonies would no longer entrust to Britain and which, as individual states, they could not perform themselves. In contrast, the British North American colonies of the mid-nineteenth century did not willingly repudiate imperial economic integration or imperial defence. Rather they found themselves substantially rejected by Britain on both counts and contrary to their wishes. At the same time their problems of economic development and defence were greatly intensified by the outstanding success which had attended the efforts of the American nation. In Confederation they created the framework of a second national state in North America on the familiar principle of fighting fire with fire. To carry the analogy further, the Canadian firemen—commonly known as the Fathers of Confederation—created a federal state and in many ways paid close attention to the American firefighting manual, the American Constitution.

A number of important steps were necessary to the creation of a

[6]The province of Canada had a net debt of $74.4 million as of June 30, 1867. Of this total $18.7 million had gone into canals and $33 million into railways. In addition the province had assumed municipal obligations of $14 million, largely associated with municipal support of railway construction. See *Report of the Royal Commission on Dominion-Provincial Relations* (Ottawa: King's Printer, 1940), Book I, p. 38.

second national, economic, and political unit in British North American territory. The first of these was the creation of a national constitution. The main outlines as well as a good deal of the detail of such a constitution were provided for in the British North America Act of 1867. Development and integration were the indispensable and inseparable economic requirements. Both rested heavily on the possibility of facilitating trade among the existing colonies or provinces. Confederation removed the tariff barr. s which had previously existed between the separate units, and the Bri. sh North America Act provided for the completion of an intercolonial railway by the national government. Further development, however, would require the exploitation of some vast new area of resources, the establishment of a new frontier of investment. Attention was thus directed even before Confederation to the prospects in Rupert's Land and in the Pacific colonies. These prospects, uncertain as they were, might be developed by the national government if ownership and transportation facilities could be assured. Thus the title of the Hudson's Bay Company to Rupert's Land and the construction of a Pacific railway were matters of serious consideration in the years before 1867.

As used in the present volume the term, national policy, without capitals, comprises collectively that group of policies and instruments which were designed to transform the British North American territories of the mid-nineteenth century into a political and economic unit.

CHAPTER TWO

Agriculture and the Investment Frontier
before Confederation

THE NATIONAL POLICY took shape over the decades before and after Confederation. The passage of the British North America Act and the federation of the four original member provinces marked the creation of the constitutional framework within which the economic and political elements of the policy could effectively be developed. If over-all objectives and certain of the methods of accomplishment were formulated with reasonable clarity by 1867, there nevertheless remained much uncertainty over what additional measures would be necessary for the achievement of these objectives. Further elaboration and substantial legal and administrative modifications were necessary before it could be said that the national policy was adequate to the task before it.

Although political leaders in the maritime colonies were the first to consider a union of the separate British colonies, it was the leaders in the province of Canada who took the initiative in proposing and promoting the larger union which evolved as Confederation. We may assume that the Fathers of Confederation envisaged measures for the enhancement of prosperity in the future which appeared reasonable to them in relation to the successes and failures of the past, not only as experienced in their own community but in that of their near-by American neighbours as well. What can be said, therefore, concerning the major economic circumstances of the St. Lawrence Valley before Confederation?

Studies of Canadian confederation ordinarily stress the difficulties which developed in the individual British North American colonies in the 1850's and 1860's.[1] This emphasis is necessary, because the federation of the provinces was regarded as the first indispensable step toward the solution of problems which the colonies could not hope to tackle successfully unless united. Confederation and the whole complex of the national policy were, however, set against a background of experience which along with all its difficulties included a considerable measure of

[1]See for example D. G. Creighton, *British North America at Confederation* (Ottawa, 1939); A. R. M. Lower, *Colony to Nation* (Toronto, 1946); and R. G. Trotter, *Canadian Federation* (Toronto, 1924).

achievement. This positive achievement, particularly as it related to economic matters, had been clearly evident in the British North American colonies in the past, it continued to be evident in the American territories near at hand, and, with proper care, it might become evident in the new Canadian confederation. For proper perspective on the national policy it is necessary, therefore, to elaborate not only the difficulties that the colonies had faced as separate units, but also to sketch the areas of economic prosperity which had existed in the past and which, it was confidently hoped, might soon be re-established.

When Canadian leaders in the 1850's and 1860's looked back, they saw clearly that earlier periods of expansive prosperity had been characterized most prominently by an abundant immigration and agricultural settlement. This was true whether the area examined was on the Canadian or the American side of the St. Lawrence. The lesson was particularly pointed up by an examination of the American economy, since prosperity and immigration persisted simultaneously there in a manner not to be mistaken for coincidence, while by 1855 prosperity and immigration had together deserted the Canadian economy. For the entire century following the Revolutionary War the most important feature of North American economic life, and indeed of the economic life of the entire Atlantic region, was the westward movement of the agricultural frontier. The British colonies on the St. Lawrence shared generously in this movement until the middle 1850's and particularly throughout the generation from 1825 to 1855. The rate of economic growth in Canada thereafter was relatively low in contrast to what it had been in the earlier period. The simultaneous decline of agricultural immigration and settlement could not escape attention.

The main facts concerning Canadian population and immigration before Confederation can be briefly outlined. At the time of the conquest of New France by Britain there were but sixty-five thousand French-Canadians on the St. Lawrence. The conquest led to the immigration of Anglo-American merchants and the purchase of French seigniories by British subjects.[2] A number of the purchasers encouraged the settlement of their seigniories by Scottish, Irish, and American immigrants. Refugee New England families settled in the Eastern Townships by 1780 and in the same year immigrant families were sent on to the posts on the Lower Lakes.[3] Loyalists from New York settled at Cataraqui (Kingston). The Loyalists who came to the St. Lawrence may have totalled six or seven thousand. Murray Bay was settled by disbanded troops before 1783. Gradually the stream of settlement increased

[2]V. C. Fowke, *Canadian Agricultural Policy* (Toronto, 1946), p. 69.
[3]*Ibid.*

in volume. The British territory on the St. Lawrence was divided into Upper and Lower Canada in 1791, and by the first decade of the nineteenth century it was estimated that their combined population was three hundred thousand, of whom from sixty to eighty thousand lived in Upper Canada.[4]

Population increase in the Canadas was thus substantial by the early part of the nineteenth century and gradually gathered momentum as the years went by. Though there were periods when immigration was small or even entirely interrupted, the total acquisition of newcomers before Confederation was very large. The population was persistently expanded by natural increase regardless of irregularity in immigration. Data which may be subject to considerable inaccuracy suggest that the population of the Canadas was 637,000 in 1825, 1,841,000 in 1851, and 2,648,000 in 1867.[5] From 1800 to 1867 it was multiplied by nine or ten. The greatest increase was in Upper Canada: in 1783 it was practically without population; in 1831 it had a population of a quarter of a million; in 1843, half a million; and in 1853, one million.[6] By 1851 its population exceeded that of Lower Canada and in 1867 the respective totals were 1,525,000 and 1,123,000.[7]

To a very considerable extent the settlement of the Canadas was but a part of the general continental movement which took place following the American Revolution. The more northerly segment occupied the Eastern Townships and flowed westward to envelop the lower Great Lakes in a steady advance toward the Mississippi Valley. The immigration of Americans, which began in volume with the Loyalists after 1780 and was sustained for years by "late loyalists and simple land seekers,"[8] was interrupted for many years by the war of 1812–14.[9] From the conclusion of the war until Confederation, immigration to Canada was largely from the British Isles. From 1815 to 1825 the number of unassisted immigrants to Canada totalled perhaps 50,000.[10] Thereafter the annual

[4]D. G. Creighton, *The Commercial Empire of the St. Lawrence, 1760–1850* (Toronto, 1937), p. 143.

[5]*Ibid.*, p. 210; and *British North America at Confederation*, p. 20.

[6]Lower, *Colony to Nation*, p. 181.

[7]Creighton, *British North America at Confederation*, p. 20. In 1851 the population totals were: Upper Canada, 951,000, and Lower Canada, 890,000. *Ibid.*

[8]See A. L. Burt, *The Old Province of Quebec* (Toronto, 1933), p. 363.

[9]Professor Lower maintains that the immigration from the United States, when eventually restored, was qualitatively significant out of proportion to its quantity.

[10]Creighton, *The Commercial Empire of the St. Lawrence*, p. 259. Helen Cowan states that the British government sponsored emigration schemes in six out of the eleven years, 1815 to 1825, and gave assistance to 7,090 emigrants. During those same six years, 65,705 unassisted emigrants left the British Isles for Canada and the maritime colonies. See Helen I. Cowan, *British Emigration to British North America, 1783–1837* (Toronto, 1928), p. 117.

totals mounted sharply. In 1827 the total was 12,648; it was 15,945 in 1829; 28,000 in 1830; 50,254 in 1831; and 52,000 in 1832.[11] Cholera epidemics were introduced by the immigrants in 1832 and 1834 and the movement was thus retarded for a time. Local political disturbances interrupted immigration between the years 1836 and 1840. With the opening of the new decade, however, the movement was restored in greatly increased proportions. It reached a peak in the catastrophic Famine Migration of 1847 when 90,000 immigrants, many of them destitute and plague-ridden, swarmed into the St. Lawrence region. Again the movement slackened, but recovered to a total of 60,000 in 1854. In 1855 the number was halved with Canadian depression and with war-created prosperity in Britain and Ireland. After the widespread economic crisis of 1857, immigration to Canada was small for several years and did not recover substantial proportions before Confederation. In so far as Canada is concerned, the years from 1825 to 1855 may be described as the years of the great migration.[12]

Canadian economic historians have not given adequate attention to migration and the agricultural settlement of the St. Lawrence Valley in the pre-Confederation period. The reason is simple. The traditional assumption is that agricultural pioneers were self-sufficient. If this assumption were correct it would follow that the pioneer farmer had no economic significance to anyone but himself. If he neither bought nor sold, it is clear that he could have had no part in an exchange economy. He would be neither borrower nor lender. He would have no economic contact with trade or industry, or, indeed, with urban life in any of its economic aspects.

An essential element of my interpretation of the origins and development of the national policy is the view that the assumption of the self-sufficiency of the pioneer farmer is incorrect. It is held to be demonstrable rather than merely arguable that the Canadian pioneer was at no time self-sufficient, that he was from the beginning of his migration and throughout his pioneer days inseparably tied in with the price system and the urban economy on a national and international basis. Most writers on the subject have protected themselves from frontal attack by describing the Canadian pioneer as "relatively," or "largely," or "in the main" self-sufficient. It is impossible to prove such statements wrong,

[11]Creighton, The Commercial Empire of the St. Lawrence, p. 259, and Lower, Colony to Nation, p. 183.

[12]Professor Creighton limits the period to the years 1825 to 1850 but the pronounced recovery of immigration after 1850 warrants the extension to 1855. Other "great migrations" in Canadian history occurred in the widely separated periods 1659 to 1673 and 1895 to 1914.

or right. Nevertheless, analyses based on such statements are misleading because they have almost universally considered the remaining margin of agricultural exchange activity to have been negligible and unworthy of evaluation. It is one of the contentions of this study that the exchange activities of the Canadian frontier settler in eastern Canada were far from negligible and that his integration into the price system did not await or depend upon his production of a staple agricultural export. This hypothesis is strongly supported by *a priori* reasoning and its validity is amply established by an examination of contemporary literature. Each support may be considered in turn.

The agricultural settlement of the St. Lawrence Valley was based on the migration of hundreds of thousands of settlers, many of them from the United States but by far the greatest part of them from the British Isles. The costly services of their transportation, whether by land or more particularly by water, had to be purchased. This is an economic fact of significance, regardless of the source of the funds. The migrants, once located on their bush lots, were unable to secure immediate subsistence therefrom and necessarily relied on their own or other capital resources for the purchase of provisions. The transportation and provisioning processes of a migration which continued for decades, even if intermittently, and which involved hundreds of thousands of persons had obvious significance for exchange and prices. Even after the pioneers were settled and had secured their first meagre crops it is difficult to imagine them as completely independent of the other segments of the economic life of the community, notwithstanding the common assumption of such independence.

The literature of that time confirms the hypothesis that the pioneer Canadian farmer was not divorced from the exchange aspects of the Canadian economy. It is obvious that the financial aspects of the transportation of prospective settlers need no confirmation. If such confirmation were needed it is readily available in the many statements in immigration and travel guides of the fares charged at different times and for different classes of accommodation. These fares were paid by the migrants themselves, by friends or relatives who had migrated earlier and had secured the funds with which to prepay passages for those who followed, by Poor Law unions, by philanthropic landlords, by charitable organizations, or by governments. Whatever the source of the funds, the transactions clearly involved the purchase of non-agricultural services by farmers or prospective farmers. The financial impact of this cash-supported agricultural demand for transportation facilities is, in fact, one of the few aspects of pioneer agriculture which Canadian economic historians have considered significant for the non-agricultural areas of the economy.[13]

Immigrants to the St. Lawrence required inland as well as ocean transportation, for on arrival at Quebec they were still hundreds of miles from their destination. The expenditures necessary for the inland portion of the journey were often half, or more than half, as much as that required for the ocean crossing. The demand for transportation services, both ocean and inland, carried with it a demand for provisions and for the commercial facilities necessary for the organized supply of such provisions.

An assumption of self-sufficiency distorts the truth about the capital requirements of frontier settlement. The pioneer settler is ordinarily verbally pictured as living with his numerous sturdy progeny in a rough-hewn log cabin, with rude, home-made furniture, tilling his "home-made" clearings with crude, home-made implements, and eating the rough but wholesome home-grown fare extracted from his land. He and his family are, in tradition, eternally clad in "home-spun," the fact that the wool from the settler's sheep was spun into yarn in the settler's cabin apparently providing convincing proof that the garments made therefrom involved the settler in no commercial transaction.

The usual idyllic picture of frontier settlement, even if accurate, would not, of course, prove that such settlement occurred without capital accumulation. Even the crudest of dwellings, furnishings, implements, clothing, and land improvement constitute real capital, and if the settler accumulated these by the painfully slow "Robinson Crusoe" methods commonly attributed to him, he was none the less engaged in a clearly recognizable task of creating capital. In modern economic terminology he was "investing" and the circumstances which induced him to do so are recognizable as "investment opportunities." But unless some commercial transactions were involved, the frontier investment would have little significance for the remainder of the economy.

In fact, however, pioneer settlement was not carried out in the bare-handed manner of common tradition. Settlers from the British Isles came from a country already far advanced in what we have since come to describe as the Industrial Revolution. Pre-eminent in the communal skills which constituted that phenomenon were those relating to the technology of textiles and of metalware, notably iron. The settlers took

[13]See, for example, H. A. Innis, "Unused Capacity as a Factor in Canadian Economic History," *Canadian Journal of Economics and Political Science*, vol. II, no. 1, Feb., 1936, pp. 1–15.

with them to the frontier the capital-goods products of the world's most advanced technical arts of the day. Many of the settlers were able to outfit themselves with their own resources. They literally took their capital equipment with them. Others were outfitted by the agencies which undertook to assist in their migration and settlement. Those who arrived in the colony with few resources—and they were in the great majority—had little choice but to work for wages in order to obtain the means of equipping themselves for bush-lot settlement. They were able to work on the farms of established settlers, in lumber camps, or on public works in Canada or in the neighbouring states. While they bought their capital equipment in the colony instead of bringing it with them, metalware and textiles at least were imported from Britain.

Contemporary sources, including the great mass of advice to emigrants published in the first half of the nineteenth century, contain many lists of articles provided, or considered to be necessary, for the person about to settle on a Canadian frontier farm. As examples of the equipment provided to settlers under assisted-settlement schemes we may refer to the provision made for two groups—the Loyalists, and the founders of Perth. In addition to free transportation, the Loyalists were given rations for a year, seed, and implements as follows: a plough, a ploughshare and coulter, a set of drag-teeth, a log-chain, an axe, a saw, a hammer, a bill-hook and a grubbing-hoe, a pair of hand-irons and a cross-cut saw, some at least of these items apportioned among several families.[14] Under a scheme for settlement in the Drummond and Bathurst townships in 1814 (the Perth founders), each settler was given the following capital goods in addition to free transportation and land: spade, adze, felling-axe, brush-hook, bill-hook, scythe, reaping-hook, pitchfork, pickaxe, nine harrow-teeth, two hoes, hammer, plane, chisel, auger, hand-saw, two gimlets, two files, a pair of hinges, one door-lock and key, nine panes of glass, one pound of putty, fourteen pounds of nails, camp-kettle, frying-pan, a blanket for each man or woman and one more for each two children. A number of items of equipment, such as pit-saws, cross-cut saws, grindstones, crow-bars and sledge-hammers, were provided on the basis of one to be shared by several settlers.[15]

Lists of equipment which prospective emigrants were advised to take with them to frontier farms in eastern Canada are informative. John

[14]C. C. James, "History of Farming" in A. Shortt and A. G. Doughty, eds., *Canada and Its Provinces* (Toronto, 1914–17), vol. XVIII, p. 564.
[15]Lorne J. Henry and Gilbert Paterson, *Pioneer Days in Ontario* (Toronto, 1938), p. 38.

MacGregor, writing from the Maritimes in the early 1830's, outlined in the following words the equipment needed by emigrants to British North America:

Farmers or labourers should bring out with them, if their means will admit, as much clothing, bedding, and linen as they may require for two or three years, a set of light cart harness, two spades, two shovels, two scythes, four sickles, four or five hoes, two pair of plough traces, the iron work of a plough and harrow of the common kind used in Scotland; the cast machinery of a corn-fan; one hand, one jack, and one jointer plane; one draw knife, six socket chisels, six gouges, one hand saw, two or three hammers, three or four augers assorted, none larger than one and a quarter inch; a dozen gimblets, a few door hinges and latches, and a small assortment of nails.[16]

A statement prepared in 1832 by the committee in charge of a scheme by landlords and parishes in Sussex for the assistance of emigration gives a list of the articles "absolutely necessary for comfort in the colony." The list, almost entirely made up of personal effects, is comprised in the statement that:

Families should take their bedding, blankets, sheets; pewter plates or wooden trenchers; knives, forks, spoons; metal cups and mugs; tea kettles and saucepans; working tools of all descriptions. The lowest outfit recommended by the parishes for their labourers [contained]: fur cap, warm great coat, Flushing jacket and trowsers, a duck frock and trowsers, a canvas frock and trowsers, two jersey frocks, four shirts, four pair of stockings, three pair of shoes, a bible and a prayer book.[17]

The capital equipment required by settlers in eastern Canada, in addition to their personal and household effects, must obviously be classed as tools rather than implements within any modern meaning of the terms. It is easy to understand how observers have dismissed as of no importance the relatively trifling equipment available to the pioneer of 1825 or 1850 in comparison with the equipment needed on an Ontario farm of 1925 or 1950. Yet this is no justification for dismissing the capital requirements of the pioneer as negligible or non-existent. The demand created by the outfitting of an individual immigrant settler was small, but the opening of the agricultural frontier in Upper Canada rested on the immigration and outfitting of scores of thousands, even hundreds of thousands of settlers. In the half-century preceding Confederation the opportunities for productive activity on the Canadian agricultural frontier provided significant investment opportunities on both sides of the Atlantic as well as in the North Atlantic shipping service.

[16]John MacGregor, British America (2nd ed., Edinburgh, 1833), vol. II, p. 532.
[17]Cowan, British Emigration to British North America, p. 183.

The agricultural immigrant to Upper Canada contributed to the creation of profitable investment opportunities as he was being outfitted at home before his departure,[18] as he purchased ocean passage, as he moved inland from Quebec and bought transportation, accommodation, and provisions on the way, and, finally, as he purchased additional equipment and provisions to take with him to his bush-lot farm. But the original settlement was only the starting-point, and similarly the original outfit of tools, personal effects, and provisions was but the starting-point in the settler's demands for capital equipment and con-sumers' goods and services. The frontier settler did not attain self-sufficiency upon arrival at his farm, or, indeed, at any time thereafter. Contemporary evidence makes it clear that his demands on the com-mercial system for capital equipment and consumers' goods persisted and became increasingly diverse in proportion as he became more and more settled.

The assumption of self-sufficiency has ordinarily rested on the absence of any staple agricultural product for export during the pioneer era. Obviously if the farmer did not sell, neither could he buy, except for the doubtful possibility that he might buy on credit with no prospect of repayment. But self-sufficiency is not the only alternative to staple production. The frontier farmer might have no single staple product and still sell a great variety of produce in substantial quantities. This, in fact, is what he did. Even before his first seed-time he reaped a crop, hard-wood ash, which brought ready cash in the local market centres where it was assembled and processed for export. Once established on his farm he fed hogs, cattle, and sheep, and thus had for sale such produce as fresh and salt pork, beef, butter, mutton, and wool. Most cereals were grown for the feeding of livestock on the farm, but wheat and peas, particularly wheat, were among the most dependable of all sources of cash.

In stressing "cash" products as we have, it is important not to fall into the error of dismissing barter transactions as of no commercial importance, or of assuming that the scarcity of money in frontier communities was a proof of self-sufficiency. Much of the produce disposed of by the pioneer was bartered, but it was none the less

[18]Miss Cowan, speaking of the assisted migrations to Quebec in the years 1832 to 1836, points out that the preparation of outfits for the emigrants "created a demand for labour in the town of Petworth, and so benefited those who re-mained at home as well as the emigrants themselves" (*ibid.*). As far as the creation of investment opportunities is concerned, the source of the funds for the settlers' outfits was immaterial: it was the *expenditure* of the funds that created the investment opportunities.

disposed of commercially and constituted effective demand for capital equipment and consumers' goods of non-agricultural origins.

Among the specific groups within the colony which created a demand for farm produce were the British garrisons, the streams of new immigrants and settlers, and the timber camps. The military garrisons were of great importance as an agricultural market in the early years after the coming of the Loyalists but declined in importance as the farm population increased. The tens of thousands of new settlers who came each year after the early 1820's created a great demand for provisions as they moved inland from Quebec and located on backwoods farms. The timber camps, or shanties, were located principally along the Ottawa River and its tributaries and created a demand for horses, hay, and oats, and for all sorts of provisions for men.[19] Although settlers in certain localities could sell their produce directly to the garrisons, the shanties, or to the immigrants and new settlers, the usual buyer of farm produce was the country merchant.[20] He was commissioned by the government to purchase grain and other produce for the garrisons. He bought for resale to immigrants and to the non-farm workers in the growing market centres. Finally, he bought for consignment to Montreal and thence to the export market.

The establishment and growth of the numerous market centres which had taken place in the St. Lawrence region by 1850 are understandable only in terms of the considerable commercial activities of pioneer settlers. The timber trade contributed to the development of these centres as the fur trade had in a much lesser degree before 1800. But the series of well-established ports which were located on the northern shore of Lake Ontario before 1850 and the scores, even hundreds, of market centres which dotted the agricultural countryside away from the lake front were supported basically by the commercial patronage of the pioneer farming communities in which they grew and prospered. Cross-roads stores,

[19]Said a commentator in the middle 1840's: "The Timber Trade . . . [of the Ottawa River] is likely to prove of permanent importance . . . ; and it gives a market at the door of the farmer for a good portion of the more bulky part of his produce,—for his horses, cattle, hay, oats, beans, peas, potatoes, and a portion of his wheat." From the *Canadian Economist*, June 6, 1846, as cited in H. A. Innis and A. R. M. Lower, eds., *Select Documents in Canadian Economic History, 1783–1885* (Toronto, 1933), p. 41.

[20]"But should there be no demand . . . [from immigrants, the settler] may carry his produce to the merchants. They will give him in exchange, broad-cloth, implements of husbandry, groceries, and every sort of article that is necessary for his family, and, perhaps, even money, at particular times. . . ." "The country merchants receive annual supplies of goods from Montreal, and send down pork, flour, staves, and potash, in return." From Howison, *Sketches of Upper Canada*, as cited in *ibid.*, pp. 46, 231.

hamlets, villages, and towns represented substantial capital formation on the agricultural frontier. The processing and merchandising equipment within them was for the most part created specifically to meet agricultural requirements.

The evidence bearing on the nature and functions of these pioneer market centres is extensive and can only be sampled here. Evidence relating to the development of Port Hope, for example, clearly indicates the agricultural orientation of the processing and merchandising facilities established at that place. Port Hope was not a backwoods settlement, but it typifies the commercial centres which grew up on the lake front to serve the agricultural hinterland. Trading posts had existed at the mouth of the Ganaraska since 1770. In 1793 Elias Smith, a United Empire Loyalist merchant, and two others brought in four families of Loyalists to settle in the district. Within the next five years they settled forty families in all. A description of the early growth of the village reveals its significant features:

The economy of the village was agricultural, each family growing or making practically all the necessities of life. There was, however, even in the earliest years of settlement some commercial life and some import trade from the United States, although most of the supplies came through the port of Montreal. The sailing ships which plied the lake anchored off shore and small tenders or flat bottomed boats went to shipboard, taking off passengers and supplies. Also, in 1793, Elias Smith sent his son, Peter, and a young man, named Collins, from Montreal to the settlement with a supply of goods and instructions to open a store. . . .

Within the first decade after its founding, Smith's Creek [later Port Hope] had a saw and grist mill, a store, a tannery, a blacksmith shop, an hotel, a distillery, and a tavern in addition to Hercheimer's trading post.

By 1817 the village had a population of 750. . . . The commercial life of the village had grown appreciably during the first quarter of the new century and, by 1826, Port Hope had four general stores; two saw and grist mills; four distilleries; one malt house; three blacksmith shops; an ashery; a wool-carding factory; a rush chair-bottom factory; a cut-nail works; a tannery; a fanning mill factory; a shoemaker's shop; an hotel; three taverns; a butcher shop; a tailor shop; a watchmaker's shop; a hatter's shop; and an auctioneer's store.[21]

This citation enumerates the capital equipment of one of Upper Canada's leading lakeshore ports and indicates clearly the extent to which such equipment was directed toward meeting the economic requirements of the pioneer agricultural community. Similar information is available concerning many of the inland market centres that developed in step with the agricultural settlement of the surrounding areas. A

[21]A. H. Richardson, *A Report on the Ganaraska Watershed* (Toronto: King's Printer, 1944), pp. 8, 10.

contemporary noted in the 1820's that Perth, for example, had "six or eight stores where you can get anything, as you or I could in Glasgow, but cast metal is very dear and crockery ware. There are four churches. . . ."[22] At the same time St. Thomas had two taverns, a mill, a church, two stores, and an academy.[23] Of Richmond it was said that, long before Bytown, "it had become a business and commercial centre of some consequence. There were at least a dozen good general stores, four breweries and two distilleries, a sawmill, gristmill, carding-mill and shops providing the general services of all the common trades, like those of waggon-maker, blacksmith, harness-maker and cooper."[24] Oakville in 1837 presented to Mrs. Jameson "the appearance of a straggling hamlet, containing a few frame and log-houses, one brick house, (the grocery store, or general shop, which in a new Canadian village is always the best house in the place) a little Methodist church . . . and an inn dignified by the name of the 'Oakville House Hotel.' " "Where there is a store, a tavern, and a church," Mrs. Jameson added, "habitations soon rise around them."[25] Oakville at this time had over three hundred inhabitants. A writer reported in 1834 concerning the settlers in the Eastern Townships that "Their markets and intercourse were chiefly in the adjoining States, with the exception of some cattle and horses, which they drove through the woods for sale in Canada, or a few articles they floated down the St. Francis when the waters were high. . . . The villages of Stanstead, Hatley, Compton, Lennoxville and Shipton and some others have assumed the neat and comfortable appearance of New England villages. Numerous stores are established, and manufactories, workshops, academies, places of worship and even printing offices, have made their appearance."[26]

It is clear that the pioneer settler relied continuously from the time of his arrival upon the commercial and processing facilities which formed the basic capital equipment of these non-agricultural centres. The hundreds of urban communities which existed in Upper Canada in the late 1840's could not have become established in areas in which the settlers were self-sufficient. The formation of real capital in these market

[22]Letter from a Scottish settler who came to the Perth settlement in 1820, as cited in Henry and Paterson, *Pioneer Days in Ontario*, pp. 225–6.

[23]*Ibid.*, p. 222.

[24]*Ibid.*, p. 226.

[25]Anna Jameson, *Winter Studies and Summer Rambles* (Toronto, reprint 1943), p. 15.

[26]From Quebec *Gazette*, quoted in Montreal *Gazette*, Oct. 16, 1834, as cited in Innis and Lower, eds., *Select Documents in Canadian Economic History, 1783–1885*, p. 35.

centres provided a substantial part of the employment and investment opportunities which gave economic significance to the immigration and agricultural settlement of the second quarter of the nineteenth century in Canada.

The above analysis suggests that there is need for a revaluation of the place of agriculture in the pre-Confederation Canadian economy and in the formulation of the national policy. By the 1840's and 1850's, agriculture in the St. Lawrence Valley was on a staple-producing basis and much of the support of the commercial interests for Confederation came from the merchants, carriers, and processors of agricultural products. Immigration and agricultural settlement, however, are processes that have commercial importance quite independent of the possibilities of staple production, and staple production, in fact, represents only the extreme stage of the possible degrees of interdependence between agriculture and commerce. Immigration and settlement in the upper St. Lawrence region were commercial as well as agricultural processes. They enormously expanded the field of investment opportunities for commercial, transportation, and manufacturing capital within and beyond the new territory. The capital equipment of the region multiplied in response to the attraction of these opportunities. Within the period from 1825 to 1855 Canadian immigration and agricultural settlement were on a scale sufficient to provide the investment vitality for the entire economy, supported, but only in lesser degree, by the investment vitality arising from the erratic expansion of the timber trade.

Before 1850 there was little realization in the St. Lawrence colony of the economic effectiveness of migration and settlement. Persons familiar with the immigration publicity of the Dominion government or of the Canadian Pacific Railway Company in the early twentieth century may find this difficult to believe. Yet as far as immigration was concerned, the old adage applied to the Canada of a century ago, that one does not miss the water till the well runs dry. The immigrants who came to the St. Lawrence before 1850 came without assistance or encouragement from the authorities of their new homeland. They were emigrants from the old land rather than immigrants to the new, if we may make such a distinction. The agent in Quebec who after 1828 supervised their quarantine and debarkation and who gave them a minimum of guidance and assistance was the "Emigrant Agent" rather than an immigration officer, and his salary was paid by the Imperial government. In years when immigration was not unusually large and when its members were persons with capital and good health, the movement was taken for granted generally and was passively welcomed by the countless persons who prospered by it. When numbers were unusually large or when, as often was the case, there was a high proportion of indigence and ill health among the newcomers, the movement gave rise to bitter protest.

There were, however, occasional signs that the economic attributes of immigration and settlement were not entirely overlooked. In 1832 the Lower Canada assembly passed a bill imposing a head tax on immigrants to create an "Emigrant Fund" to aid in the settlement of indigent migrants. The merchants in Lower Canada attempted to block the bill in council and, failing in that, they petitioned the Governor to disallow it, arguing in part that it would divert immigration to the United States. The Upper Canada legislature and other groups petitioned against the bill. One editor in Upper Canada expressed his opposition to it in a view which was not at that time widely voiced, but which twenty years later was to become an integral part of the working philosophy of Canadian government and was so to remain for three-quarters of a century. He said: "We have for some time past enjoyed a valuable influx of Emigrants, or as it might be termed, a valuable import trade of the nerves and sinews of prosperity—of the true wealth of the country, an industrious peasantry. . . ."[27]

Not until the 1850's did Canadian governmental policy come to be squarely based on a full realization of the significance of immigration and agricultural settlement for the well-being of the entire economy. This realization, latent, it might be said, for a generation, was finally brought to conscious formulation by the comparative inadequacies of immigration and agricultural settlement after 1850. This is not to suggest that these processes ended suddenly or that the 1850's were years of stagnation in Canada. Immigration was heavy in the earlier part of the decade and settlement and improvement of land progressed rapidly, particularly in the western portion of the province. The population of Upper Canada increased by 47 per cent and that of Lower Canada by 25 per cent from 1851 to 1861. The Hungry Forties were past. The Reciprocity Treaty, in conjunction with eastern American industrialization and urban development, provided greatly increased markets for Canadian agricultural exports such as cattle, barley, rye, and oats. The Crimean War forced the price of wheat up to a peak of $2.40 per bushel in Toronto. Canal improvement was temporarily at an end but heavy investment in railway construction, the building and improvement of roads, and the creation and expansion of market-centre capital

[27]Cited in Creighton, *The Commercial Empire of the St. Lawrence*, pp. 274–5.

equipment provided employment possibilities and a high level of community income.[28]

Commencing in the early 1850's, while prospects for further agricultural expansion were still comparatively promising, leaders in the province of Canada pursued two projects relating to the agricultural frontier. One was the attempt to speed the expansion of the agricultural frontier within the province by the promotion of immigration and by the adoption of an active settlement policy. The other was a continuation of the long-maintained but hitherto unsuccessful effort to bring the Canadian commercial economy into effective contact with the American agricultural frontier. Its chief feature was the costly and seemingly futile construction of the Grand Trunk Railway from the American boundary on the west to Montreal, and thence to an ice-free port on the American seaboard.

The effort to extend the agricultural frontier within the province of Canada involved primarily the establishment and maintenance of the first official agency in British North America for the promotion of immigration. This was the Bureau of Agriculture established in the provincial government in 1852. The head of the Bureau, ex officio Minister of Agriculture, was directed by law to grant and record patents, to have charge of the census and statistical returns, and to "institute inquiries and collect useful facts and statistics relating to the Agricultural interests of the Province, and to adopt measures for disseminating or publishing the same in such manner and form as he may find best adapted to promote improvement within the Province, and to encourage immigration from other countries."[29] Working under these directions, the personnel of the Bureau—which became the provincial Department of Agriculture in 1862 and the federal Department of Agriculture following Confederation—directed their energies almost exclusively to the vigorous promotion of immigration from overseas.[30] The other main feature of the attempt to extend the Canadian agricultural frontier consisted in

[28]Professor Pentland has published a stimulating exploratory study of certain aspects of the processes of capital formation in relation to Canadian economic development. Speaking of the imports of capital he says: "The capital injections [imports] that spurred development may be put roundly, then, at $10 million from 1827 to 1837, $15 million in the forties, $100 million in the fifties, and $200 million from 1867 to 1875. With each dose, the effect lessened proportionally as the economy grew and broadened. The peak effect was experienced in the middle fifties." See his "The Role of Capital in Canadian Economic Development before 1875," *Canadian Journal of Economics and Political Science*, vol. XVI, no. 4, Nov., 1950, pp. 457–74, particularly p. 465.

[29]*Statutes of Canada*, 16 Vic., c. 11 (1852), s. 6.

[30]For an analysis of the activities of the Bureau of Agriculture see Fowke, *Canadian Agricultural Policy*, pp. 112–16 and 121–5.

modifying provincial land policy in order to make local settlement attractive to newcomers in competition with the apparently superior attraction of settlement prospects in mid-western American states.[31] Colonization roads were opened through the Ottawa-Huron tract and one-hundred-acre lots along them were offered free of charge to bona fide settlers.

Attempts to expand the Canadian agricultural frontier in the 1850's failed as completely as did the efforts to establish commercial contact with the American agricultural frontier. Immigrants entering at Quebec averaged 10,000 a year from 1858 to 1860 and 20,000 a year from 1861 to 1865. These figures were barely comparable with the yearly totals of a generation before when no assistance had been extended to immigrants. The most discouraging contrast, however, was with the persistent buoyancy of American immigration and settlement. By 1860 New York was receiving seven immigrants to every one arriving in Quebec. Free grants and colonization roads on the Canadian frontier were inadequate to divert the European immigrants from settling in the American Middle West where land was obtainable only at a price. Instead of proving a land of opportunity for scores of thousands of new settlers each year, Canada was, in fact, unable to retain her native sons.

It was in this context of economic frustration that the national policy was gradually evolved.

[31]For a discussion of these efforts see *ibid.*, pp. 125–34.

CHAPTER THREE

The Western Territories and the National Policy in Its Formative Years

THE EMERGENCE OF THE NATIONAL POLICY represented a major change in the views of St. Lawrence commercial interests concerning the location of the Canadian frontier. Successive generations of resident and non-resident merchants, from the earliest establishment of European trading-posts on the St. Lawrence River, had refused to recognize any limit to their commercial domain within the confines of the continent. Until the beginning of the nineteenth century the commerce had been in furs and the investment frontier had been a fur trader's frontier, but after the Revolutionary War, and particularly after the beginning of the nine-teenth century, the westward expansion of agricultural settlement on a continental basis completely overshadowed the fur trade as a source of investment opportunities for labour and capital. St. Lawrence merchants abandoned the fur trade and directed their efforts toward the servicing of the new frontier. The political boundary which was gradually projected across the continent by the various piecemeal settlements effected be-tween the American and British governments was wholly unrealistic as far as migration and agricultural settlement were concerned. The expan-sive forces stemmed particularly from the American seaboard, but pioneer settlement spread westward in almost complete disregard of the political boundary, peopling the empty spaces of British North America as readily as those lying immediately to the west of the American coastal colonies. As in the fur trade, so in the agricultural trade, commerce was continental. Merchants on the St. Lawrence, at Montreal, regarded the St. Lawrence as the natural trade route to the inland frontier, wherever that frontier might lie. Merchants at New York, on the other hand, con-sidered the Hudson-Mohawk river system to be the natural entry to the frontier.

The American Revolution introduced nationalism to North America. Originally it was merely a defensive nationalism. The main problems for the new nation were, first, to establish the best possible geographic boundaries between British and American territories and, second, to seize as many as possible of those commercial advantages which the

American colonies had enjoyed as colonial members of the first British Empire. Improvement of the Hudson-Mohawk river system by means of the Erie Canal, completed in 1825, gave the Americans undisputed superiority over the St. Lawrence river system in the commerce of the interior and, in fact, threatened the St. Lawrence merchants in their own territory to the north of the lower Great Lakes. By the time the St. Lawrence canal system was complete—in the late 1840's—the American merchants were trading with the agricultural frontier over a railway system which joined the Mississippi Valley with the Atlantic seaboard. The Americans approached the Oregon boundary dispute in the 1840's in the spirit suggested by their slogan, "Fifty-four forty or fight."[1] The establishment of the 49th parallel of latitude as the boundary through the Oregon territory to the Pacific Coast in 1846 represented a partial American defeat, but a defeat regarded by many as purely temporary. Simultaneously, American publicists rallied to the banner of "Manifest Destiny," the concept defined in the years of its early effectiveness as "the right of our manifest destiny to overspread and to possess the whole of the continent which Providence has given us for the development of the great experiment of liberty and federated self-government entrusted to us."[2] By 1850 the early defensive nationalism of the United States had given way to an aggressive economic and political philosophy which fully sanctioned territorial aggrandizement within continental limits.

Continental ambitions were not, however, exclusively American, as we have seen. Aggressiveness had by no means been lacking in the conduct of the commercial pursuits of the St. Lawrence merchants. The construction of the Grand Trunk Railway in the 1850's may nevertheless be regarded as the last attempt at economic aggression on the part of St. Lawrence commercial interests. It became increasingly clear that the American commercial system, and the American manufacturing system as well, were more than a match for any rivals. By 1860 it was apparent that the St. Lawrence commercial system had no chance of sharing in the agricultural trade of the American frontier. If the St. Lawrence merchant group were to look to an expanding agricultural frontier for profitable employment, it would of necessity be one of their own creation. Throughout the second quarter of the nineteenth century

[1]The southern limits of Russian claims on the Pacific Coast had been fixed in 1824 and 1825 at 54° 40′. Thus the American claim, if accepted, would have excluded Britain from the Coast entirely. See F. W. Howay, W. N. Sage, and H. F. Angus, *British Columbia and the United States* (Toronto, 1942), p. 120.

[2]From a statement by John L. O'Sullivan in the New York *Morning News*, Dec. 27, 1845, as cited by Albert K. Weinberg in *Manifest Destiny: A Study of Nationalist Expansionism in American History* (Baltimore, 1935), p. 145.

the Canadian agricultural frontier had expanded sufficiently without effort or encouragement to provide adequate business opportunities for all levels and branches of the St. Lawrence merchant class. In the 1850's, intensive encouragement to immigration and settlement were quite incapable of advancing the Canadian frontier to any appreciable extent. Such advance was, in fact, effectively halted at the margins of the Precambrian Shield. The British or Canadian counterpart of the hundreds of fertile miles through which the American frontier advanced before reaching the Mississippi was the Precambrian rock and forest which stretched for a thousand miles around the northern shores of the upper Great Lakes. Beyond the Shield the territory was unknown to any but fur traders and a handful of Red River settlers. Here even the promise which might derive from favourable American conditions failed.

Between the Mississippi and the Rocky Mountains lay the Great American Desert, so prominent in American written and spoken communication in the 1850's.[3] Convincing proof of the existence and extent of the desert was the fact that the Oregon settlers refused to settle in or near the heart of the vast continental plain. Instead they trekked beyond its western limits and traversed its two thousand miles at an estimated cost of seventeen lives per mile.[4] This nice statistical computation concerning mortality was undoubtedly made much later. The horrors of the Oregon Trail were nevertheless common knowledge and confirmed the general belief that the continental plain constituted a great and effective barrier to agricultural advance.

American publicists had no interest in indicating how far the desert extended into the territories of the Hudson's Bay Company beyond the international boundary and neither Canadians nor Britons had such information ready to hand. Indicative of a general lack of knowledge about western territories and, at the same time, of a recognition of the need for developmental prospects beyond the Precambrian Shield were the inquiries and explorations directed toward these territories by Canadian and British authorities in the late 1850's. In 1857 both the Canadian Legislature and the British House of Commons appointed select committees to hear evidence concerning the rights and administrative

3"The tradition of the Great American Desert was at its height in the decade between 1850 and 1860. Most of the textbooks showing it were published about 1850 and were in use through the decade. . . ." "The fiction of the Great American Desert was founded by the first explorers, was confirmed by scientific investigators and military reports, and was popularized by travelers and newspapers." Walter Prescott Webb, *The Great Plains* (Boston, 1931), pp. 159, 153.

4*Ibid.*, p. 149.

efficacy of the Hudson's Bay Company, and at the same time to consider the possibilities for agriculture and settlement of the Northwest Territories.[5] Both committees recommended against the continued possession of the territories by the Company. The position of the Company in the western territories was constantly under attack after the middle of the century.

In the same year, 1857, the Colonial Office and the Legislature of the province of Canada dispatched exploratory parties to the northwest to determine the productive capabilities of the region in terms of its timber and its mineral and agricultural resources, and to discover what possibility there might be of opening the region for immigration and settlement by means of transportation facilities which might be established through British territory. Captain Palliser, in charge of the British expedition, carried on his explorations till 1860. He reported that the Great American Desert extended into British territory, but only far enough to create a southern triangle of infertile land outside which, between the Shield and the mountains, there was a great fertile area suitable for settlement. He reported further, however, that with the international boundary established at the 49th parallel of latitude, a railway to the west coast entirely through British possessions was out of the question. Professor Hind, a member of the Canadian expedition of 1857, returned in 1858 to explore the valleys of the Saskatchewan and Assiniboine rivers. He found many handicaps to settlement but reported that lack of markets rather than climate or soil would prove the chief barrier to agricultural development.

Long before any definite conclusions could be drawn, three facts became apparent. First was the obvious fact that no new development could be expected in the West as long as the Hudson's Bay Company was left in its position of absolute proprietorship. The Company's primary and almost exclusive interest in the central plains continued to lie in the fur trade as it had always done. The treatment accorded the Red River settlement by the Company left no doubt that active encouragement of immigration and settlement was not to be expected of it.

[5]In 1857 the government of the province of Canada sent Chief Justice Draper as a delegate to appear before the Select Committee of the British House of Commons investigating the affairs of the Hudson's Bay Company. He was instructed by his government "to urge the expediency of marking out the limits [of the Northwest Territories], and so protecting the frontier of lands above Lake Superior, about the Red River, and thence to the Pacific, as effectually to secure them against violent seizure or irregular settlement, until the advancing tide of emigrants from Canada and the United Kingdom . . . fairly flow into them and occupy them as subjects of the Queen on behalf of the British Empire." Minute of Council, Feb. 18, 1857, in Public Archives of Canada, State Book R, 223 ff., as quoted in R. G. Trotter, *Canadian Federation* (Toronto, 1924), pp. 236–7.

The second fact to become apparent was that western development could not be fostered and guided from the East without adequate transportation between the regions, though the most suitable kind and location of such facilities remained uncertain. By the 1850's the steam railway was the accepted new device for land transportation. It was the established instrument in the development of the American western frontier, but the tremendous geographic disadvantages of the British territories made it extremely doubtful whether a railway could seriously be contemplated.

The third fact which developed from the preliminary examination in the 1850's was that any attempt to develop western prospects would be a task so great as to be far beyond the capabilities of any one of the British American colonies or provinces as currently constituted. The Americans with much greater resources had not attempted western expansion on the basis of the separate efforts of the individual colonies or states, but had made it one of the primary purposes of their new national government. The example could not be ignored by Canadians.

Canadians, in fact, had an analogy if not a parallel even closer at hand. The province of Canada was itself a united territory, a constitutional unit created by legislative union in 1841. One of the major reasons for this union had been the inability of either of the colonies which formed part of the St. Lawrence region to undertake the developmental tasks required in that region. The union had made possible the completion of the St. Lawrence canal system in the 1840's and was essential to the construction of the Grand Trunk and other railways in the 1850's. The united province would be inadequate to the colossal task involved in the projection of transportation facilities to the western territories; but the province of Canada was, after all, only a partial union of the eastern British provinces. A union of all of them would be the smallest possible aggregation of strength with which to approach the problems of western development.

The necessity for adequate transportation facilities was apparent from the earliest days of the emergence of the national policy. If attention was directed exclusively to the proposal to integrate the eastern territories, an intercolonial railway would be indispensable, for the natural waterway which connected the St. Lawrence and the Maritimes was obviously ineffective. If union with the western territories was to be contemplated as well, communications would have to be established from the eastern to the western limits of the proposed political unit.

Discoveries of gold on the Fraser River in the 1850's and in the Caribou in the 1860's attracted attention to the British territories on the Pacific Coast. In 1856 the President of the Canadian Executive

Council, *ex officio* Minister of Agriculture, remarked that Canada's western boundary should be on the Pacific.[6] In 1858 Alexander Morris published in a pamphlet entitled *Nova Britannia* the text of a lecture which he gave in Montreal. The pamphlet was widely circulated. Morris' theme was federal union which, Professor Trotter says, "was strongly allied in his mind with the question of the acquisition of the North-West, still monopolized by the Hudson's Bay Company. . . . [His pamphlet constituted] a glowing account of the growth and possibilities of British North America from the Atlantic to Vancouver Island, and foretold the day when this wide domain would be united politically."[7]

Proposals for a Canadian Pacific railway had been advanced as early as 1849[8] but the development of interest in the Pacific coastal regions in the 1850's imparted reality to ideas which had originally appeared to be but the product of idle fancy. Both Captain Palliser for the British government and S. J. Dawson[9] for the Canadian government were charged with the examination of transportation possibilities when they were dispatched on their western explorations.

[6]V. C. Fowke, *Canadian Agricultural Policy* (Toronto, 1946), p. 139.

[7]Trotter, *Canadian Federation*, p. 31.

[8]In 1849 Carmichael Smyth recommended the construction of a railway from Halifax to the Pacific Coast by means of convict labour. He estimated that the line would need to be 4,000 miles in length and would cost £24,000 per mile. He suggested that the railway be managed by a board of fifteen members on which Great Britain, the Hudson's Bay Company, Nova Scotia, New Brunswick, and the United Province of Canada would each have a representation of three. See S. J. McLean, "National Highways Overland" in A. Shortt and A. G. Doughty, eds., *Canada and Its Provinces* (Toronto, 1914–17), vol. X, pp. 419–20. In 1851 Joseph Howe told an audience in Halifax that many of them would live to hear the whistle of the steam engine in the Rocky Mountains and would be able to go from Halifax to the Pacific in five or six days. See G. P. de T. Glazebrook, *A History of Transportation in Canada* (Toronto, 1938), p. 236. In 1851 Allan MacDonnell of Toronto urged the construction of a railway from the Great Lakes to the Pacific. In advocacy of his proposal he published a pamphlet entitled, *A Railroad from Lake Superior to the Pacific: The Shortest, Cheapest and Safest Communication for Europe with Asia* (n.p., n.d.). See McLean, "National Highways Overland," p. 420. MacDonnell and his group had a bill introduced into the Canadian Assembly in 1851 to incorporate the Lake Superior and Pacific Railroad Company. This bill was reported on unfavourably by committee on the grounds of lack of feasibility. Similar bills were defeated in 1853 and 1856, but in 1858 a group secured a charter for the North-West Transportation, Navigation and Railway Company. Glazebrook, *A History of Transportation in Canada.*

[9]A member, with Professor H. Y. Hind, of the party sent out by the government of the province of Canada in 1857 under the direction of a Mr. Gladman to explore a transportation route from Lake Superior to the Red River settlement. Two parties were sent in 1858, the first, as has been mentioned, under the direction of Professor Hind to explore the valleys of the Saskatchewan and Assiniboine rivers, and the second under the direction of S. J. Dawson to continue the survey of a possible transportation route from Lake Superior to Fort Garry.

A striking illustration of the way in which the various elements of the national policy began to fall into place is provided by the activities of E. W. Watkin. In 1860 Mr. Watkin, a member of the British House of Commons, accepted the task of salvaging the finances of the newly constructed Grand Trunk Railway for its British owners. He exacted the pledge that the Imperial government would give consideration to a scheme for the union of the British American colonies to be "followed eventually by a railway from coast to coast."[10] Without these conditions, he was convinced, the Grand Trunk could never be placed on a paying basis. His views on the matter are clearly stated in a letter which he wrote to a friend in November, 1860. In his letter he said in part:

This line [the Grand Trunk] both as regards its length, the character of its work, and its alliances with third parties, is both too extensive and too expensive, for the Canada of today; and left, as it is, dependent mainly upon the development of population and industry on its own line, and upon the increase of the traffic of the west, it cannot be expected, for years to come, to emancipate itself thoroughly from the load of obligations connected with it. . . . [The way out], however, to my mind, lies through the extension of railway communication to the Pacific. . . .

Try to realize, . . . assuming physical obstacles overcome, a main through Railway, of which the first thousand miles belong to the Grand Trunk Company, from the shores of the Atlantic to those of the Pacific, made just within—as regards the northwestern and unexplored district—the corn-growing [wheat] latitude. The result to this Empire would be beyond calculation; it would be something, in fact, to distinguish the age itself; and the doing of it would make the fortune of the Grand Trunk.[11]

Watkin was instrumental in arranging the intercolonial conference at Quebec in 1864 to revive the intercolonial railway project and to further the union of the provinces which to him appeared essential. Since the Hudson's Bay Company still held the heart of the continent and might block the plans for a Pacific railway, Watkin organized a group in London to buy control of the Company from its shareholders.

In Watkin's activities as sketched in the preceding paragraphs we may see important elements of the national policy taking shape. Here in integrated form, and directed toward the solution of the financial problems of one of Canada's most powerful economic interests, was action contributing to at least four of the segments of national policy: disposal of the Hudson's Bay Company's barrier to the West; a Pacific railway; union of the British territories; and development of the West from the Red River settlement to the Pacific Coast.

[10]See A. R. M. Lower, *Colony to Nation* (Toronto, 1946), p. 313.
[11]As quoted in Trotter, *Canadian Federation*, pp. 179–81.

Reference to Mr. Watkin and his Canadian activities emphasizes the fact that the pressures working toward the national policy in Canada arose by no means exclusively in the North American continent. The possibility of creating a unified British possession stretching from the Atlantic to the Pacific, and the certainty that to obtain such a possession meant building a transcontinental railway carried important implications for Britain in relation to the pattern and prospect of world trade. These implications can be fully appreciated only by reference to the age-old quest of Europeans for a short route to the Indies, a quest which narrowed in the Western world to the search for the western sea.

The land masses of the Western Hemisphere had first appeared to European discoverers as a tremendous barrier lying athwart a possible water route to the Indies. Throughout the centuries which followed the voyages of Columbus, Europeans never abandoned the search for the Northwest Passage. By the end of the eighteenth century it had become apparent that such a passage, entirely by open water, did not exist. The western continents would have to be circumvented, as they could be by way of hazardous Cape Horn, or portaged, as they might be at the Isthmus of Panama or even at other places where the extent of land was immensely greater. Not till railway technology was well advanced, however, as it was by the middle of the nineteenth century, was there a serious possibility of a trade route from Europe to the Orient across rather than around the Americas. By the same date, Britain had established new commercial contacts with far-eastern countries, notably China. Neither the Suez nor the Panama canal had yet been constructed.

Under these circumstances it became apparent that transcontinental North American railways were potential and essential allies of any group or region desirous of competing effectively in the renewed world rivalry for the far-eastern trade. In the letter written in 1860 and cited above, E. W. Watkin advocated the extension of the Grand Trunk Railway to the Pacific, and commented in partial justification: "Try for one moment to realize China opened to British commerce; Japan also opened; the new gold fields in our own territory on the extreme west, and California, also within reach; India, our Australian Colonies—all our Eastern Empire, in fact, material and moral, and dependent (as at present it too much is) upon an overland communication through a foreign state."[12]

[12]*Ibid.* The United States Senate Committee on Pacific Railroads, speaking in 1869 of the Northern Pacific Railway then under construction, made reference to Asiatic trade in a manner typical of the railway literature of the Confederation and post-Confederation period. "From China (Canton) to Liverpool," the Committee said, "it is 1500 miles nearer by the 49th parallel of latitude, than by way of San Francisco and New York. This advantage in securing the overland trade from Asia

Professor Lower, speaking of Watkin's designs for western development and a western railway, says, "To Watkin and the London group the North West Territories came in as valuable in themselves, but still more valuable as a route to Asia."[13]

Since the term "national policy" was not used contemporaneously to symbolize the purposes that were shaping up in the British North American territories of the 1850's and 1860's, it is useless to search the literature of the day under that heading to learn of the place of the western territories in the formulation of the policy in question. The union of the provinces, however, was of central importance in the formative period of the national policy, not only as a measure of importance in itself but also because union had to be effected before progress could be made toward working out the other essential features of the policy. In 1865 the Legislature of the province of Canada debated the proposals for Confederation as contained in the resolutions of the Quebec Conference of 1864. The views expressed by the various speakers in these debates suggest the expectations associated with the project for the creation of a national state. Ordinarily spoken of as the *Confederation Debates*,[14] they may equally well be thought of as representing the first of a series of debates on national policy.

Our emphasis upon the economic aspects of the national policy should not be taken to imply that the national purposes were at any time exclusively economic. The diversity of intentions involved is made clear in the *Confederation Debates*. In addition to a persistent emphasis on the defensive aspects of the proposed union, considerable weight was attached to its potential contribution to the solution of the political difficulties which existed in the United Province of Canada. In introducing a major speech in the Debates of 1865, Hon. Thomas D'Arcy McGee, Minister of Agriculture, reviewed the arguments advanced by preceding speakers and added: "But the motives to such a comprehensive change as we propose must be mixed motives—partly commercial, partly military, and partly political; and I shall go over a few—not strained, or simulated—motives which are entertained by many people of all these prov-

---

will not be thrown away by the English, unless it is taken away by our first building the North Pacific road, establishing mercantile agencies at Puget Sound, fixing mercantile capital there, and getting possession on land, and on ocean, of all machinery of the new commerce between Asia and Europe." *Report of the Senate Committee on Pacific Railroads*, as cited in Howay, Sage, and Angus, *British Columbia and the United States*, p. 199.

[13]*Colony to Nation*, p. 313.

[14]*Parliamentary Debates on the Subject of the Confederation of the British North American Provinces*, 3rd session, 8th provincial Parliament of Canada (Quebec, 1865), hereafter referred to as the *Confederation Debates*.

inces, and are rather of a social, or, strictly speaking, political than of a financial kind."[15]

On the economic side of the matter most of the emphasis was placed, in the *Confederation Debates*, upon the elaboration of benefits which might be anticipated from the expansion of interprovincial trade following the construction of the Intercolonial Railway and the removal of interprovincial tariff walls. A great deal was said about the economic advantages of a diversity of resources such as would be created by a union of the maritime and St. Lawrence areas. Complementary resources, it was argued, made for growth and stability. Galt said: ". . . it is in the diversity of employment that security is found against those sad reverses to which every country, depending mainly on one branch of industry, must always be liable. . . . We may therefore rejoice that, in the proposed Union of the British North American Provinces, we shall obtain some security against those providential reverses to which, as long as we are dependent on one branch of industry as a purely agricultural country, we must always remain exposed."[16]

A reading of certain sections of the *Confederation Debates* would leave one with the impression that the new nation was to unite within its boundaries the maritime provinces and the province of Canada, but nothing further. Or one might gather from the same sections that, although the new national boundaries might eventually extend beyond these limits, the economic advantages of the union would in any case arise solely from the expansion of trade and commerce between the St. Lawrence and the eastern seaboard regions. A number of speakers, however, including several of the outstanding leaders of the day, saw the union of the eastern provinces simply as the starting-point in the creation of a political and economic unit which would stretch from the Atlantic to the Pacific and which would provide economic advantages in all its parts. There is ample support in the *Debates* for Professor Trotter's remark that, "The notion . . . that the acquisition of the North-West by the new Dominion was a fortunate afterthought, has little basis in fact."[17]

George Brown was the most forceful and persistent of the advocates of western development as he had been for more than a decade before. Very early in the *Debates* Holton raised the question of the place of the western territories in the federation scheme. Directing his remarks to Brown, he "desired some information as to the position of the North-

[15]*Ibid.*, p. 128.
[16]*Ibid.*, p. 63.
[17]Trotter, *Canadian Federation*, p. 222.

West question on which the President of the Council [Brown] has always taken strong grounds, maintaining that Canada had a territorial right extending over all that region. He took it for granted that the President of the Council still maintained his position, but he wished to know from him authoritatively the manner in which the Government proposed to deal with the question."[18] Brown replied that "Hon. Mr. Holton had done no injustice to him in supposing he held now precisely the same sentiments on the North-West question he formerly did. He believed it of vast importance that that region should be brought within the limits of civilization, and vigorous measures had been taken to ascertain what could be done with that view."[19]

In a later speech in the *Debates*, Brown directed attention to the agricultural resources of the various parts of the British territories. Dealing first with the untilled areas of central Canada, he added: "And if the mind stretches from the western bounds of civilization through those great north-western regions, which we hope ere long will be ours, to the eastern slope of the Rocky Mountains, what vast sources of wealth to the fur trader, the miner, the gold hunter and the agriculturist, lie there ready to be developed. (Hear, hear.) . . ."[20]

The reference to the wealth awaiting the fur trader in the West was partly a doffing of the hat to history and partly a reference to one of the few certain resources of the western region. Mostly, however, it was a sharp thrust at the foundations of French-Canadian opposition to the federal plan for national development. In a more direct attempt to undermine this opposition Brown said in part:

It has always appeared to me that the opening up of the North-West ought to be one of the most cherished projects of my honorable friends from Lower Canada. . . . it has always struck me that the French Canadian people have cause to look back with pride to the bold and successful part they played in the adventures of those days . . . in their prosecution of the North-West fur trade. (Hear, hear.) Well may they look forward with anxiety to the realization of this part of our scheme, in confident hope that the great north-western traffic shall be once more opened up to the hardy French Canadian traders and *voyageurs*. (Hear, hear.)[21]

But there were others in Lower Canada besides the *voyageurs* who would recall the glories of the western fur trade. These, the Anglo-Canadian merchants, would be more deeply impressed by a figure in sterling, by a reminder that the western fur trade was once carried on by way of the

[18]*Confederation Debates*, p. 17.
[19]*Ibid.*, p. 18.
[20]*Ibid.*, p. 98.
[21]*Ibid.*, pp. 103–4.

St. Lawrence. For them there would be significance in the confident assertion that western trade, whatever it might eventually prove to be, could again be restored to its "natural route" by way of the St. Lawrence river system. Brown went on:

Last year furs to the value of £280,000 stg. ($1,400,000) were carried from that territory by the Hudson's Bay Company—smuggled off through the ice-bound regions of James' Bay, that the pretence of the barrenness of the country and the difficulty of conveying merchandise by the natural route of the St. Lawrence may be kept up a little longer. Sir, the carrying of merchandise into that country, and bringing down the bales of pelts ought to be ours, and must ere long be ours, as in the days of yore, and when the fertile plains of that great Saskatchewan territory are opened up for settlement and cultivation, I am confident that it will not only add immensely to our annual agricultural products, but bring us sources of mineral and other wealth on which at present we do not reckon.

McGiverin, member for Lincoln, pointed up the trade route argument as sharply as could be by putting into words the parallel of American experience which was undoubtedly often present in the minds of Canadians who thought of federation as a step toward western development. He said: "What then may we not expect our great North-West to become? If we had it opened up, Canada would be the carriers of its produce, as the Middle States are the carriers of the Western States, and the manufacturers of its goods as the Eastern States are now the manufacturers of the goods consumed by the west. . . ."[22]

Keeping in mind the economic experience of Canada and the United States throughout the decades before 1860, it is not surprising to find immigration and agricultural settlement advanced strongly in the *Confederation Debates* as among the prime purposes of the national plan and particularly of western development. On this point Brown said:

I go for a union of the provinces, because it will give a new start to immigration into our country. . . . On this question of immigration turns, in my opinion, the whole future success of this great scheme which we are now discussing. Why, Sir, there is hardly a political or financial or social problem suggested by this union that does not find its best solution in a large influx of immigration. . . . And this question of immigration naturally brings me to the great subject of the North-West territories. (Hear, hear.) The resolutions

[22]*Ibid.*, p. 470. McGiverin foresaw more clearly than most of his contemporaries the agricultural possibilities of the West. "If the North-West contains land, as I believe it does," he said, "equal to almost any on this continent, it should be placed in precisely the same position as regards Canada that the Western States occupy in relation to the Eastern. I believe we should endeavor to develope [*sic*] a great grain-producing district; for whatever may be said, there is not any appreciable quantity of grain-producing land in the hands of the Government not now under cultivation in Canada, for the benefit of our increasing population." *Ibid.*, p. 469.

before us recognize the immediate necessity of those great territories being brought within the Confederation and opened up for settlement.

If we wish our country to progress [as the United States], we should not leave a single stone unturned to attract the tide of emigration in this direction; and I know no better method of securing that result, than the gathering into one of these five provinces and presenting ourselves to the world in the advantageous light which, when united, we would occupy.[23]

McGiverin uttered the sad reminder that it was not only a question of encouraging immigration, but also a matter of preventing the emigration of native sons: "It is a melancholy fact that for the want of such a country, our youth seek homes in a foreign land. . . . If we had that country open to them, to say nothing of the foreign immigration it would attract, it would afford homes for a large population from amongst ourselves now absorbed in the Western States."[24]

As Minister of Finance, Galt was particularly sensitive to the opposition to the national project which rested on fears of greatly increased financial burdens. His argument was that expenses for such things as adequate defence and public works were unbearable for states which were small and weak but could be easily borne by states with adequate resources:

. . . with limited resources and undeveloped territory it might be impossible for any small country to undertake the necessary outlay [for defence and public works]. . . . In this view let us look at the immense extent of territory that stretches away west of Upper Canada. The reason why we have not been able to assume possession of that territory and open it up to the industry of the youth of this country . . . is because the resources of Canada . . . have been inadequate for the development of this great district. Now, one of the resolutions of the scheme before the House refers to this same question, and I believe that one of the first acts of the General Government of the United Provinces will be to enter into public obligations for the purpose of opening up and developing that vast region, and of making it a source of strength instead of a burden to us and to the Mother Country also. (Hear, hear.) . . .[25]

Opposition critics attacked various features of the national plan in the debates of 1865 but not all of their criticism was designed to impede it. Some members, it was clear, felt that the plan as outlined in the Quebec resolutions was not sufficiently comprehensive or that it did not provide adequate guarantee for the early completion of the total design. Particular concern was expressed that western development, while clearly contemplated in the scheme, might be indefinitely postponed. Honour-

[23]*Ibid.*, pp. 102–4.
[24]*Ibid.*, pp. 469–70.
[25]*Ibid.*, pp. 70–1.

able Mr. Skead, for example, challenged the government as to the seriousness of its intentions concerning the development of the West:

I would even say that the scheme of the delegates to the Quebec Conference does not go far enough. I contend that, instead of merely taking in the provinces to the east of us, the scheme should have embraced British Columbia and the whole of the territory to the west. An honorable friend near me says that will come in good time. But I am afraid that some Downing Street or other influence may prevent it. (Cries of "no, no!") I should like to see the Pacific as the western boundary of this young Confederation, in the same way as the Atlantic is its eastern limit, so that we should have one country stretching from ocean to ocean. (Hear, hear.) . . . [After completion of the Intercolonial Railway] we shall then be ready to commence the railroad to British Columbia, and the improvement of the Ottawa river to the upper lakes, and the navvies and others who have been employed on these works will find employment on the road leading to the Pacific, and will ultimately become settlers in the great Red River country. . . .[26]

At one stage in the debate, Brown felt impelled to answer charges that "while the Intercolonial Railroad had been made an absolute condition of the compact, the opening up of the Great West and the enlargement of our canals have been left in doubt."[27] Calling to his support resolutions 68 and 69 of the Quebec Conference, he asserted: "Now, Sir, nothing can be more unjust than this [charge]. The Confederation is . . . clearly committed to the carrying out of both these enterprises."

Since our concern at this point is with the very question involved in the interchange noted above, it will be worth our while to quote the resolutions on which Brown relied for his assertions about the West.

68. The General Government shall secure, without delay, the completion of the Intercolonial Railway from Rivière du Loup, through New Brunswick, to Truro in Nova Scotia.

69. The communications with the North-Western Territory, and the improvements required for the development of the trade of the Great West with the seaboard, are regarded by this Conference as subjects of the highest importance to the Federated Provinces, and shall be prosecuted at the earliest possible period that the state of the finances will permit.[28]

Brown was undoubtedly right in stating that the Quebec resolutions were designed to commit the proposed federal government to both the Intercolonial Railway and to the development of the West. There nevertheless remained a clear differentiation in the degree of urgency of the projects. The government, caught between the pressures of those who felt that it was moving too rapidly and of those who were sure that it

[26]*Ibid.*, pp. 243–4.
[27]*Ibid.*, p. 103.
[28]*Ibid.*

was moving too slowly, necessarily exercised a considerable measure of caution in any statements which purported to indicate the time when the various parts of the national plan might be implemented. The *Confederation Debates* make clear two points at least concerning the scope and timing of the plan: first, it was the firm intention that the new nation would ultimately extend from the Atlantic to the Pacific with all parts developed in between; second, for the immediate future there was contemplated only the union of the eastern provinces and the construction of a railway to link the St. Lawrence and the maritime regions. As for the establishment of communications between the eastern provinces and the West, and the formulation and prosecution of an effective plan for the development of the western territories, these would come in time, but when, no one was prepared to say precisely. Brown, as staunch an advocate as could be found of ultimate continent-wide national development, refused to accede to pressure from the opposition to guarantee that western development should proceed simultaneously with the construction of the Intercolonial Railway.[29] A statement which he made early in the *Confederation Debates* sums up the views of the government concerning the national plan and the place of Confederation in its ultimate accomplishment: "Sir, the whole great ends of this Confederation may not be realized in the lifetime of many who now hear me. We imagine not that such a structure can be built in a month or in a year. What we propose now is but to lay the foundations of the structure—to set in motion the governmental machinery that will one day, we trust, extend from the Atlantic to the Pacific."[30]

[29]*Ibid.*, pp. 475–6. He said that the government was sincere in its determination to proceed with western development "at the earliest possible moment."

[30]*Ibid.*, p. 86. Not more than a month earlier (Jan. 3, 1865), the *Globe* had stressed the legal uncertainties surrounding the western territories: "The peculiar position of the North-West question formed the reason why the Convention at Quebec could not deal more definitely with it. No member of that body had any authority to speak for the North-West. It was one thing to agree definitely to secure the Intercolonial Railway through provinces represented in the Convention and a totally different thing to speak of a road to pass through a territory for which the Convention had no sort of claim to speak." As quoted in D. G. Creighton, *British North America at Confederation* (Ottawa, 1939), p. 42.

CHAPTER FOUR

velopment and Integration of the
National Policy

THE BRITISH NORTH AMERICA ACT of 1867 established the political constitution, the first step needed for the elaboration and implementation of the national policy, and created the federal government, the major instrument by means of which the plan was to be carried out. It is necessary at this point to indicate the nature of the constitutional provisions made in the British North America Act for the furtherance of the national project and to examine the development and integration of the main non-constitutional elements of the national policy, that is, those relating to railways, immigration, settlement and land policy, and protective tariffs.

As clearly foreshadowed in the Quebec resolutions and in the *Confederation Debates*, the British North America Act federated only part of the British North American territories but provided for the ultimate union of the whole. The immediate objective was to unite the maritime colonies and the province of Canada, but only Nova Scotia, New Brunswick, and the province of Canada were joined, the latter to be divided into Ontario and Quebec. The act, however, made specific provision for the later entry into the federation of Newfoundland, Prince Edward Island, and British Columbia, and of Rupert's Land and the Northwest Territories.[1] The construction of the Intercolonial Railway, to run from the St. Lawrence River to Halifax, was made an immediate obligation of the new federal government, the act declaring it to be (s. 145) "essential to the Consolidation of the Union of British North America, and to the Assent thereto of Nova Scotia, and New Brunswick." The act made no mention of a Pacific railway or of the establishment of communications with the western territories.

[1]Rupert's Land and the northwestern territory were united with Canada by Order in Council in 1870. The Rupert's Land Act and the Rupert's Land (Loan) Act, both of 1868, had provided for the purchase of the Hudson's Bay Company's rights. The colony of British Columbia was admitted to the Canadian federation by Order in Council in 1871, and Prince Edward Island was admitted similarly in 1873. Newfoundland was admitted in 1949 by a British North America Act, 1949, which provided for the modification of several sections of the act of 1867 in its application to the new province. See H. McD. Clokie, *Canadian Government and Politics* (new and revised ed., Toronto, 1950), Appendix A, p. 328 n.

Contrary to the first intentions of a number of the leaders of the Confederation movement, the constitution as formulated, discussed, and finally adopted, provided for a federal rather than a legislative union of the provinces. This meant the creation of a new hierarchy of governmental bodies. The parliamentary system, already well established in the provincial governments of British North America, could readily be adapted to the new constitution, but the federal principle, as new to the provinces as it was to British tradition, required the solution of the problem of how to allocate powers and responsibilities between the federal or national government and the provincial governments. It is well recognized that the draftsmen of the British North America Act wasted few words in outlining the modifications in the parliamentary aspects of government made necessary by the new constitutional arrangements, but went to considerable length to spell out their intentions concerning the division of responsibilities between the national and the provincial governments. With all their care they did not escape ambiguity, and, partly because of their care, they left a legacy of rigidity not infrequently associated with written constitutional instruments.

The present analysis does not require elaborate discussion of the division of powers as provided for in the British North America Act or of the constitutional problems that have arisen. The national government was created in order to undertake those responsibilities which were inescapable and at the same time beyond the abilities of the individual colonies, responsibilities which may be summarized under the headings of national defence, national development, and the public debt. The first obligation was taken care of by assigning to the federal government exclusive control over militia, military service, and defence. The responsibility for national development involved constructing the Intercolonial Railway and, in particular, controlling commercial matters as set out in some detail in section 91. Herein was established exclusive federal jurisdiction over the regulation of trade and commerce, and over currency and coinage, banking and the issue of paper money, weights and measures, interest, legal tender, bankruptcy, patents and copyrights, and a number of allied matters. The Dominion government assumed the debts of the provinces on their entry to the federation. Since the bulk of the debts accumulated by the provinces by 1867 had been incurred for developmental purposes, for the construction of canals, railways, and roads, their assumption by the federal government amounted to a retroactive assumption of the burdens of national development.

The provinces were left with the responsibilities which in the Canada of a century ago were regarded as matters of purely local concern. These included education and a considerable range of activities which today would come under the heading of public welfare. The provinces were also left in control of the public lands within their respective territories on the assumption, apparently, that no prospect for substantial economic development remained therein. The provinces and the Dominion were given concurrent jurisdiction over agriculture and immigration with a provision for the superiority of Dominion jurisdiction in case of conflict.

A consideration of the development of other features of the national policy may well start with railway policy. The point may be emphasized that the national policy was a product of the railway era. The coming of the railway created many of the problems for which the national policy was designed to provide the solution, but it also made possible the transcontinental design of the national policy and was inevitably one of its essential features. The unequal impact of the railway upon the different sections of the continental economy was disruptive of competitive positions and impelled the construction of additional lines designed to restore the balance. The St. Lawrence canal system, completed in the 1840's, might well have placed the St. Lawrence trade route in a position of competitive competence vis à vis the Hudson-Mohawk river system had the latter remained on the barge-canal basis provided by the Erie Canal. But American railway construction rendered canals obsolete and precipitated the costly and ineffective Grand Trunk Railway venture.

At every stage of the evolution of the Canadian national design, American economic expansion and railway construction established the competitive goals and set the pace for Canadian planners, particularly in regard to the West. The views concerning the western territories that were expressed in the *Confederation Debates* of 1865 made it possible to assess in the preceding chapter the place of the West in the national design as that design had evolved to that time: western development was regarded as of the utmost importance in the long run, but quite beyond the range of immediate requirements and possibilities. However, within three years of Confederation, when the British Columbia delegates came to Ottawa with instructions to request a Pacific wagon road as one of the conditions of union, the Canadian government proposed a railway instead, and gave commitments providing for early commencement and completion. Development of the West was obviously regarded with a much greater sense of urgency in 1870 than it had been even half a decade before. This change in sentiment requires explanation.

The American concept of Manifest Destiny as formulated in the 1840's had related specifically to the Pacific Coast region and the Oregon boundary dispute, although worded in continental terms. With the British-American agreement in 1846 that the 49th parallel of latitude would serve as the boundary from the foothills to the coast, as had already been agreed upon for the territory between the Great Lakes and the Rocky Mountains, and with a commitment on the part of the Hudson's Bay Company to promote the agricultural development of Vancouver Island,[2] it appeared that American expansion and territorial claims had been at least temporarily checked.

Whatever the immediate effect of the settlement of the Oregon boundary dispute, there was no significant alteration in the speed or direction of American economic development and effective territorial occupation. The western agricultural frontier came closer and closer to the international boundary in the neighbourhood of the Red River settlement. The population of Minnesota increased from 6,000 to 172,000 in the years between 1850 and 1860 and the territory became a state of the Union in 1858. After a difficult struggle, the Red River settlers established for themselves the right of freedom of trade in defiance of the monopoly claims of the Hudson's Bay Company, and, by the 1850's, commercial relations were firmly established between the settlement and St. Paul, a commercial outpost of the United States. River boats and annual convoys of Red River carts provided transportation.

That commercial penetration was likely to be but a forerunner of eventual territorial occupation was strongly suggested by certain incidents in the American Congress. A bill providing for the absorption of all British North American territories into the American Union was introduced into the House of Representatives in 1866.[3] It was defeated, but in 1867 Senator Ramsey of Minnesota moved that the Committee on Foreign Relations investigate the desirability of a treaty between the United States and Canada (sic) which, among other things, would provide for the annexation of all territories in North America west of the 90th meridian; he stated that there was currently a move in Canada to extend Canadian boundaries to the Pacific Coast, but that the Red River and British Columbian settlers would prefer to join the United States. Although Senator Ramsey's resolution was tabled for the time being, it was brought forward again in 1868 and passed with minor amendments. The Senate Committee on Pacific Railroads, to which the resolution was then referred, apparently regarded the matter as one which need not be dealt with hastily since it could have but one outcome and that favourable to the American desire for territorial expansion.

The greatest and most persistent of the competitive pressures shaping

[2]Vancouver Island was granted to the Hudson's Bay Company in January, 1849, under an agreement requiring the Company to establish settlements of British subjects thereon within five years. This was done with the deliberate intention of forestalling peaceful penetration by the Americans. See F. W. Howay, W. N. Sage, and H. F. Angus, *British Columbia and the United States* (Toronto, 1942), p. 137.
[3]*Ibid.*, p. 196.

Canadian policy, however, lay in the western and northern expansion of the American railroad network. The completion of the Union Pacific and Central Pacific link between the Mississippi region and the Pacific Coast in 1869 was an event of the greatest significance in terms of continental development in North America and of revolutionary importance in terms of world trade routes. It represented the first transcontinental railway link to be constructed anywhere in the world. It was hundreds of miles distant from the borders of the new Canadian nation; but the Northern Pacific, chartered in 1864, was a different matter. The latter railway was projected to link Lake Superior with Puget Sound and thus to bring the northern central plains as well as the northern mountain and coastal regions into close connection with the eastern American commercial and manufacturing base.

The Northern Pacific—as were the other American railways—was a private organization but was regarded by the American Congress as an instrument of national, continental policy. In 1868 and 1869 the American Senate Committee on Pacific Railroads urged its early completion as an effective instrument for the American economic conquest and eventual occupation of the British plains area. This is made clear in their report:

The line of the North Pacific runs for 1500 miles near the British possessions and when built will drain the agricultural products of the rich Saskatchewan and Red River Districts east of the mountains, and the gold country of the Fraser, Thompson and Kootenay Rivers west of the mountains. . . . The opening by us first of a North Pacific Railroad seals the destiny of the British possessions west of the 91st meridian. They will become so strongly Americanized in interests and feelings that they will in effect be severed from the new Dominion and the question of their annexation will be but a question of time.[4]

The threat to the Canadian national project which lay in the American railway expansion of the 1860's was quite apparent to Canadian governmental leaders. This, along with the defensive strategy which the threat called forth, is evident in the following quotation:

Early in 1870, C. J. Brydges, the general manager of the Grand Trunk, reported to Macdonald a conversation with Governor Smith of Vermont, then President of the Northern Pacific: "I am quite satisfied from the way Smith talks to me, that there is some political action at the bottom of this . . . to prevent your getting control for Canada of the Hudson's Bay Territory." "It is quite evident to me," Macdonald replied, "not only from this conversation, but from advices from Washington, that the United States Government are resolved to do all they can, short of war, to get possession of the

[4]As quoted in *ibid.*, p. 199.

western territory, and we must take immediate and vigorous steps to counter-act them. One of the first things is to show unmistakably our resolve to build the Pacific railway."[5]

It is not surprising to find Macdonald and other eastern leaders con-cluding on the basis of information such as this that a railway from Canada to the Pacific Coast must be built and built without delay. It is not so commonly known that defensive impulses prompted at least one proposal for the construction of a railway to be built *from* the Pacific to the central plains. Walter Moberly, appointed Assistant Surveyor General of British Columbia in 1864, decided as a result of explorations which he conducted in the years 1864 to 1866 that a good railway could be built from the Pacific Coast through the Rockies to the prairie country. He visited the western states in 1867 to see how American western railroads were progressing. He noted the progress of the Union Pacific and Central Pacific, well on the way to a western junction, but he was particularly interested in the Northern Pacific with its projected route much closer to the British Columbian border. He said he was anxious "to ascertain the probable effect the building of this latter railway would have in drawing away Canadian trade into American channels, in order that my proposed Canadian railway should be prepared to meet such an emergency by branch lines properly located for that purpose."[6]

The British government had not pressed for the union of British Colum-bia with the Dominion until after 1869 when negotiations were concluded for the transfer from the Hudson's Bay Company to the Crown, and thence to the Dominion government, of the title to Rupert's Land. This accomplished, the British government favoured the inclusion of the coastal province since, as Lord Granville said, one government common to all regions could more readily undertake "the establishment of a British line of communication between the Atlantic and Pacific Oceans."[7] Union with British Columbia and the pledge to construct a Pacific rail-way were logically embodied in the one agreement. In the words of a group of British Columbian scholars: "The railway was a national under-taking, essential to the building up of the Dominion from sea to sea. It was also to be a link in an imperial chain, a vital part of the 'all red route' which was to bind together the various portions of the far-flung Empire. The United States Ambassador in London informed the Foreign

[5]Brydges' comments and Macdonald's reply quoted from Joseph Pope, *Correspondence of Sir John Macdonald* (New York, 1921), pp. 124–5. See Chester Martin, *"Dominion Lands" Policy*, Canadian Frontiers of Settlement, ed. W. A. Mackintosh and W. L. G. Joerg, vol. II, part II (Toronto, 1938), pp. 225–6.
[6]Howay, Sage, and Angus, *British Columbia and the United States*, p. 233.
[7]*Ibid.*, p. 195.

Secretary that the United States would not regard its completion with favour. The construction of the railway was, in short, an act of high policy. . . ."[8]

The commitment to build the Pacific railway within ten years alarmed many of the supporters, as well as the political opponents, of the government. To allay their fears a resolution was introduced in the House and, after considerable debate and modification, was accepted in the following form: "That the railway should be constructed and worked as a private enterprise, and not by the Government, and that the public aid to be given to the enterprise should consist of such liberal grants of land, and such subsidy in money or other aid not increasing the present rate of taxation."[9]

The Liberals—then in opposition—attempted to have this resolution incorporated into the terms of union between the Dominion and British Columbia, but were overruled. The claim was in fact frequently urged in later years that the commitment of the Dominion government to British Columbia was conditional upon this resolution. Such a claim could not be supported in view of the facts of the case. Nevertheless this resolution provided an approach to the Pacific railway which was adopted by the federal government thereafter, an approach modified in the name of expediency by both political parties but never repudiated by either. After 1871 it was taken for granted in Canadian public life that the Pacific railway was to be built by private enterprise aided by grants of land, subsidies, and other assistance to an extent which would occasion no increase in the rate of federal taxation.

Under the terms of union with British Columbia, the Dominion government undertook the commencement of the Pacific railway within two years from the date of union (July 20, 1871), and assured its completion within ten years. The government lost little time in starting on the project, although the time originally allowed for completion was to pass with little achievement. In the session of 1872 the government passed an act[10] to provide for the Canadian Pacific railway and to incorporate two railway companies,[11] each one expecting the award of the Pacific franchise. Failing to secure the amalgamation of the two companies, the government incorporated a third, the Canadian Pacific Railway Company,[12] but still

[8]*Ibid.*, p. 237.

[9]Canada, *House of Commons Debates*, April 11, 1871, as cited by Edward Blake in *ibid.*, April 15, 1880, p. 1426.

[10]*An Act Respecting the Canadian Pacific Railway*, 35 Vic., c. 71 (assented to June 14, 1872).

[11]The Inter-Oceanic Railway Company, incorporated by statute, 35 Vic., c. 72; and the Canada Pacific Railway Company, incorporated by statute, 35 Vic., c. 73.

[12]Incorporated by letters patent under date of Feb. 5, 1873.

not the company of that name today. This company failed to meet its initial commitments and was involved along with the Conservative administration in the Pacific Scandal of 1873.

The Liberal administration which replaced the Conservatives following the election of 1873 made new legislative provision for the construction of the Pacific railway in an act[13] which contained alternative provisions for public or private construction of the railway. The Liberals incorporated no Pacific railway company. Throughout their term of office, which ended in 1878, they carried on varied and considerable preliminary work, including construction, toward the accomplishment of the project, and all by means of government employees or by contractors under government contract. The most important section constructed between 1873 and 1878 was that between Winnipeg (or St. Boniface) and Emerson on the international boundary. This line, on the east side of the Red River, and known currently and in later years as the "Pembina branch," linked up with the St. Paul and Pacific Railway[14] at Emerson. This was the "end of steel" at that time for the network of railways which was spreading rapidly throughout the American Middle West. The Pembina branch was put into operation in 1879 and provided the first rail communications between western Canada and the outside world.

The new Conservative government, elected in 1878, proceeded with survey work and contract construction on the Pacific railway for approximately two years before it concluded an agreement with the syndicate which assumed the project and carried it to completion. This syndicate, incorporated by Dominion statute[15] in 1881 as the Canadian Pacific Railway Company, opened rail communication between Montreal and Port Moody on the Pacific Coast in June, 1886. One or two points about the development of the Company which are frequently overlooked may be mentioned.[16]

[13]By *An Act to Provide for the Construction of the Canadian Pacific Railway*, 37 Vic., c. 14 (assented to May 26, 1874).

[14]The St. Paul and Pacific was bought out of receivership in 1879 by the St. Paul, Minneapolis and Manitoba, a company organized for the purpose by George Stephen, Donald Smith, R. B. Angus, J. J. Hill, and one or two others. This company, with 565 miles of completed road in 1879 and with a land grant of 2½ million acres, expanded a decade or so later into the Great Northern Railway Company. See O. D. Skelton, *The Railway Builders* (Toronto, 1916), pp. 131–9; H. A. Innis, *A History of the Canadian Pacific Railway* (Toronto, 1923), p. 93 n.

[15]*An Act Respecting the Canadian Pacific Railway*, 44 Vic., c. 1 (assented to Feb. 15, 1881).

[16]For standard references see Innis, *A History of the Canadian Pacific Railway*; Skelton, *The Railway Builders*; G. P. de T. Glazebrook, *A History of Transportation in Canada* (Toronto, 1938). For a shorter reference see V. C. Fowke, *The Purposes, Origins, and Development of the Canadian Pacific Railway Company as a Corporate Entity: Evidence in the Matter of an Application by the Canadian*

The first point to be noted is that the Pacific railway, so essential to the implementation of the national policy, was not necessarily a transcontinental railway: it needed only to extend from the Pacific Coast to a junction with the railways of the St. Lawrence area so as to link the East and West on an all-Canadian basis. The Intercolonial Railway provided the link with the east coast. Apart from the Grand Trunk Railway, which reached the western boundary of Ontario at Sarnia, the railway system of the central provinces reached northwesterly beyond Ottawa where, in the late 1870's, the Canada Central Railway was approaching Lake Nipissing. The legislation which incorporated the Canadian Pacific Railway Company in 1881 defined the Canadian Pacific Railway as a line to run from a point near the south end of Lake Nipissing to the Pacific Coast. This was, in fact, the definition which had been embodied in the Canadian Pacific Railway acts of 1872 and 1874, and the contractual obligations of the Canadian Pacific Railway Company related only to the line as thus defined. The charter instruments of 1881, of course, provided the Company with ample authority to expand the contractual main line—from Callander, near Lake Nipissing, to Port Moody on the Pacific Coast—into a transcontinental and intercontinental system of transportation and communication.

Another point to be emphasized is that the construction of the Canadian Pacific Railway was in fact a joint achievement of the Canadian Pacific Railway Company and the Canadian government. The extensive assistance given to the Company by the government included $25 million and 25 million acres of agricultural land "fairly fit for settlement"; land for right of way, sidings, and structures; substantial exemptions from taxes and duties; and a twenty-year guarantee against certain competitive construction in the West.[17] In addition the government agreed to complete those sections of the railway which it had already under construction in 1881, and to transfer them to the Company on completion, without equipment, but without cost to the Company. These were the Lake Superior section from Fort William to Winnipeg, and the western section from Kamloops to Port Moody. The Pembina branch was also to be conveyed to the Company. The portions of the railway thus constructed by the government and transferred to the Canadian

Pacific Railway Company for the Establishment of a Rate Base and Rate of Return before the Board of Transport Commissioners for Canada, January 5, 1953 (Regina, 1953).

[17]The monopoly clause was cancelled in 1888 by agreement between the Company and the federal government, the government guaranteeing the interest on a bond issue of the Company to the amount of $15 million in consideration of the cancellation.

Pacific Railway Company totalled 713 miles and cost the government $37,785,320.[18]

The national aspects of the Pacific railway project and of the Canadian Pacific Railway Company are nowhere more clearly indicated than in the prolonged discussions regarding the location of the Company's main line. The chief point at issue was the location of the route between eastern Canada and Fort Garry. The commitment made by the Dominion government in 1870 left the route for the Pacific railway unspecified and undecided. In general there were three possible ways of linking the railway system of eastern Canada with the Red River. First, the Grand Trunk Railway which ran from Montreal to Sarnia could be extended westward to Fort Garry without encountering insuperable obstacles. Construction might be extended through American territory to the south of the Lakes; or running rights might be secured over American railroads, which by 1870 were already spreading rapidly throughout the Middle West, and necessary extensions added to complete the route. Reference has already been made to the fact that Grand Trunk officials proposed an extension of this sort in 1860 as the only possible solution to the financial problems of the Grand Trunk system.[19] A second possibility was to build from the southeast through Canadian territory to Sault Ste Marie and to build, or secure running rights, through American territory to the south of Lake Superior. The third alternative was the construction of an all-Canadian route which meant facing the task of building through the nearly impassable area of the Shield to the north of Lake Huron and Lake Superior.

Britain's sympathy with the Confederacy during the Civil War left her North American territories, in effect, in the position of a military hostage. When the military threat abated late in the sixties, the threat of economic aggression and occupation assumed unprecedented menace, symbolized by American railway construction throughout the West. The construction of a western Canadian railway carried an increasingly heavy responsibility for defence, if not on the military, none the less on the economic, level. Regardless of the difficulties of the all-Canadian route and the added financial burden which its construction and use were certain to involve, there could be no serious consideration of a Canadian western railway by any other than a Canadian route, and by 1871 it had been decided that the Pacific railway should be built exclusively through

[18]*Report of the Royal Commission to Inquire into Railways and Transportation in Canada* [*Drayton-Acworth Report*] (Ottawa: King's Printer, 1917), p. xvi.

[19]See the letter written by E. W. Watkin in 1860 quoted on p. 31 of chap. III above.

Canadian territory.[20] The decision was never altered. In 1880, when the Conservatives were casting about to find a group of interests willing and financially able to push the Pacific railway to completion, they approached Sir Henry Tyler, President of the Grand Trunk Railway, in London. He agreed to undertake the project on condition that the plan to construct a link to the north of the Lakes be dropped in favour of one to the south of the lakes. Tupper and Macdonald refused to accept this condition and no agreement could be reached. When the proposed contract with the Canadian Pacific syndicate was debated in the Canadian House of Commons in 1880 the location of the eastern link was one of the points of opposition criticism. Their argument was that although eventually an all-Canadian line would probably have to be constructed, meanwhile there would be great economy of effort in building only to the Sault in eastern Canada and in linking up with an American railway from that point to a point on the boundary south of Fort Garry. Skelton describes the reaction to this argument: "The government urged the necessity of building at once an all-Canadian route, regardless of the added expense. . . . there was much buncombe in the flag-waving answer made [to the opposition case]. Yet, on the whole, so necessary to national unity was an unbroken road, so hard a country was this to make one, that it was best to err on the side of safety. The political interests at stake warranted some risk of money loss."[21]

Among the central figures in the Canadian Pacific Railway syndicate were George Stephen, president of the Bank of Montreal, Donald Smith of the Hudson's Bay Company, interested anonymously, and J. J. Hill; these three had also been the central figures in the St. Paul, Minneapolis and Manitoba Railway which was linked with the Pembina branch railway in 1879. The profits which the promoters of the St. Paul, Minneapolis and Manitoba Railway secured on the American frontier were adequate to ensure their interest in further expansion of their holdings. A prime condition of the interest of at least some of the members of the Canadian Pacific Railway syndicate was the possibility that their Amer-

[20]Captain Palliser had reported that the choice of the 49th parallel as the Canadian-American boundary had removed any possibility of building a transcontinental railway exclusively through British territory. Sandford Fleming was appointed engineer in chief by the federal government in 1871 and was assigned the task of finding an all-Canadian route. "It had been decided," states O. D. Skelton in *The Railway Builders*, p. 118, "in order to hold the balance even between Montreal and Toronto, to make the proposed Pacific road begin at some angle of Lake Nipissing." Sandford Fleming found a practicable all-Canadian route from that point to the coast early in his surveys and at least as early as 1876 was able to indicate the location of such a route in some detail. See *ibid.*, map opposite p. 118.

[21]*Ibid.*, pp. 145–6.

ican lines might serve as a link between eastern and western Canada during an indefinite postponement of the construction of the north-shore section of the Pacific railway. In 1882, J. J. Hill left the directorate of the Canadian Pacific Railway Company and sold his stock in that enterprise because the Company was determined to construct the Lake Superior section forthwith.

The construction of the Lake Superior section in 1883 ended both hopes and fears that it might be postponed indefinitely if not abandoned entirely. It is a tribute to the persistence of the Grand Trunk Railway and Toronto interests, however, that the whole issue was reopened after 1900 when the federal government adopted the principle that additional transcontinental railway facilities were essential in the national interest. The Grand Trunk offered to participate in the new project on the basis that it be assisted in constructing an extension to the Canadian West from Chicago, to which point it had extended its facilities by 1880. The Company did, in fact, share in the railway expansion of the early twentieth century in a position of the utmost importance, but not on the basis of an eastern link via the United States.[22]

While the alternative routes for the Pacific railway in the east were *either* Canadian *or* American, all possible routes through the mountains of British Columbia were exclusively Canadian. It was a question of farther north or south. The northern pass, the Yellowhead, was by a considerable margin the easiest pass that could be found and it would permit a railway to run northwesterly from Selkirk by way of the fertile park belt. It did not provide the shortest route to the settlements at the mouth of the Fraser. Passes farther south would permit more direct rail routes to the coast, but were immensely more difficult; furthermore, any railway using them would necessarily extend for hundreds of miles through the semi-arid plains instead of through the park belt.

The route of the Pacific railway as originally planned and surveyed was the northern route by way of the Yellowhead pass. However, the route finally adopted by the Canadian Pacific Railway Company was the southern route, across the semi-arid plains (in which the Company refused to select any substantial part of its land grant except for an irrigation area between Medicine Hat and Calgary),[23] and through the

---

[22]See below, pp. 55–6.

[23]Practically the last act in the settlement of the land-grant agreement between the Dominion government and the Canadian Pacific Railway Company took place in 1903, when the Company agreed to accept 2,900 thousand acres of land in a compact block between Medicine Hat and Calgary. In this block the Company developed a massive irrigation project. James B. Hedges, *The Federal Railway*

extremely difficult Kicking Horse Pass. The comparative directness of this route was one factor in its favour. More important was its proximity to the American boundary and the fact that this proximity offered defensive possibilities against the threat of economic invasion by American railways. We have already called attention to the completion of the Union Pacific railway in 1869. The Northern Pacific had reached Bismarck, North Dakota, on its way from the Great Lakes to Puget Sound when its progress was arrested by the financial collapse of 1873. By 1880 this railway was again under construction and J. J. Hill was projecting his railway, later reorganized as the Great Northern, to the Pacific Coast by a route even closer to the international boundary. If the Northern Pacific was regarded by the Americans as rendering inevitable the economic and political union of the Canadian West with the United States, how much more perilous would be the position of the western territories following the construction of the Great Northern? A Canadian railway following the northern arc of the park belt would be of little use in averting the American threat. The projection of the Canadian Pacific Railway through the heart of the Canadian portion of the Great American Desert may be described as a reverse adaptation of the scorched-earth method of defence.

The Pacific and the Atlantic railways (the latter the Intercolonial) were the costliest of the prerequisite instruments of the national policy. Expediency rather than principle dictated whether these railways would be built by the state or by private companies substantially financed by state benefactions and guarantees. The Intercolonial Railway was built as a government undertaking as provided for in the British North America Act. The construction of the Pacific railway was attempted both publicly and privately and was finally completed by the Canadian Pacific Railway Company, a private syndicate. But this Company was well fortified by government grants of cash, land, completed line, an exceptional degree of freedom from rate control, and a monopoly clause. Legally, the Intercolonial Railway and the Canadian Pacific Railway emerged as distinct and sharply contrasting types of institutions—the one, state; the other, private. Functionally, they began and continued as substantially similar institutions—agencies of the state designed for the furtherance of the national policy.

Land Subsidy Policy of Canada (Cambridge, Mass., 1934), pp. 52–67. In 1935 the farmers on the irrigated properties in the district formed the Eastern Irrigation District, which was deeded all the land and irrigation structures of the Company in the area and a grant of $300 thousand in cash besides. See *Report of the Royal Commission on the South Saskatchewan River Project* (Ottawa: Queen's Printer, 1952), pp. 140–1.

The construction of the Crow's Nest Pass branch[24] by the Canadian Pacific Railway Company in the 1890's was an extension of the Canadian railway system clearly dictated by the national policy. The matter at stake for the Company and for the nation was the commercial mastery of a segment of Canadian territory lying close to the American boundary and endowed with a measure of economic promise which had but recently been displayed. Discoveries of gold in the Kootenay Valley of British Columbia in the 1860's and again in the 1880's prompted intensive prospecting in the area which, in turn, led to the discovery of rich silver and base-metal properties and the development of a considerable mining activity. The north–south location of the mountain ranges and intervening valleys and the proximity of the mineral areas to the American boundary made the locality much more readily accessible to American territory and to American water and rail communications than to Canadian. Base-metal extraction involved smelting, a process which required coking coal. Smelters were already located on the American Pacific Coast and in the inland states. American railways were built northward into Canadian territory and the ores moved by waterway and over these railways to the American smelting centres. The whole area became a hinterland to the northwestern states.

Such a development constituted a major breach in the national policy which required that economic development in Canadian territory should be integrated into the Canadian economy rather than the American. Vancouver and Victoria protested the threat to their commercial interests. The Legislative Assembly of the Northwest Territories favoured the establishment of railway connections which would assure a market in the new mining areas of British Columbia for the agricultural products of the Canadian prairies in competition with similar products from the state of Washington.

The Canadian Pacific Railway Company was vitally interested in assuring that the supply and outward movement of the produce of this region should move over its lines instead of being drawn southward to move over American transcontinental railways. There were two problems: to improve transportation facilities within the region, and to join the region with the main line of the Canadian transcontinental railway. The first required the construction of rail connections between the Columbia and the Kootenay valleys. The second could be solved by

[24]For a fuller discussion of the Crow's Nest Pass railway see "Historical Analysis of the Crow's Nest Pass Agreement and Rates," Part I of *Crow's Nest Pass Rates on Grain and Grain Products: Joint Submission of the Governments of Alberta, Saskatchewan and Manitoba to the Royal Commission on Transportation* (Ottawa, Jan., 1950).

joining the Arrow lakes to the transcontinental by a short north–south branch line, or by building westward from a point on the main line on the plains, through the Crow's Nest Pass to the Kootenay mining territories. From 1889 to 1896 the Company was active in securing control of local transportation within the Kootenay-Columbia territory and in establishing rail connections between their main line and the Upper Arrow Lake. Meanwhile a variety of factors indicated the desirability of a rail line between the Kootenay and the main line of the Canadian Pacific Railway by way of Crow's Nest Pass. Such a line would provide an all-rail entry to the region from Canadian territory while the link at Revelstoke was rail-lake-and-rail, with the resultant necessity for trans-shipment of freight. Rail connections were already constructed from the prairies toward the Crow's Nest Pass in a line built in 1885 from Dunmore, Alberta, to Lethbridge for the development of the coal properties in the Lethbridge region. Finally, great resources of coking coal were found at Crow's Nest Pass. In 1892 the Dominion government made statutory provision for the extension of the Dunmore–Lethbridge line through the Crow's Nest Pass to Hope, British Columbia. The Canadian Pacific Railway Company proposed the absorption and extension of this line, and entered into a lease and option of purchase toward that end in June, 1893.

In 1897 the Canadian Pacific Railway Company entered into a subsidy and rate-control agreement with the Dominion government for the construction of the Crow's Nest Pass railway.[25] The government agreed to grant to the Company a subsidy of $11 thousand per mile, but not exceeding a total of $3,630 thousand, for the construction of a railway from Lethbridge, Alberta, through the Crow's Nest Pass to Nelson, British Columbia. The Company, on its part, agreed to build the railway to specifications as to standard, location, and time; it agreed to permit the control of certain of its rates; and it agreed to introduce specific reductions in certain other rates.[26] The national interest was of such impelling importance that the federal political parties of the day could disagree over the details of the subsidy proposed but not at all over the basic question of the subsidy itself. The Crow's Nest Pass line was completed to Kootenay Lake by October, 1898. Slowly but surely throughout the succeeding years the Canadian Pacific Railway Company

[25]The agreement was entered into under a federal statute, *Statutes of Canada*, 60–61 Vic., c. 5 (1897).

[26]The Company agreed to reduce rates on grain and flour moving eastward to Fort William and beyond, by 3 cents per hundred pounds, and on agricultural implements and a specific list of settlers' equipment and supplies moving into the West, by specific percentages, typically 10 per cent. The reductions were to be effective in perpetuity.

established effective control over movements of commodities to, from, and within southeastern British Columbia.

The addition of a second and a third transcontinental Canadian railway in the decade preceding the First World War represented an important extension of the transportation features of the national policy. The mistakes in governmental policy which resulted in such an unwarranted expansion of the Canadian railway milage have long been apparent to students of the national scene. The decisions which led to the final unfortunate result and the circumstances in which those decisions were made can readily be recounted. The decisions themselves can only partially be explained.[27]

At the turn of the century there were three important groups of private railway interests in Canada, two of them important in terms of achievement and one in terms of ambition. The Canadian Pacific Railway Company by 1900 had a transcontinental rail and communication system with a substantial complement of branch lines within and without Canada. The Company was also in possession of inland and overseas steamship facilities and a well-rounded structure of ancillary services. The Grand Trunk Railway—the second of the three—extended from the ice-free Atlantic port of Portland, Maine, via Montreal and Toronto to Sarnia, and by 1880 had extended westward to Chicago. This Company also had a considerable network of branch lines which, in this case, were entirely in the St. Lawrence region and extended northward to North Bay. Despite repeated rebuffs, the Grand Trunk Railway Company had not abandoned its early ambition to participate in the servicing of western Canadian development and in the carriage of transcontinental traffic. In the West, William Mackenzie and Donald Mann, partners in the business of railway contracting, had organized the Canadian Northern Railway Company—the third of the trio—with an accumulation of prairie milage, and by 1902 had built eastward to Port Arthur. The management of the Canadian Northern was anxious to expand further east as well as west.

With the Grand Trunk desirous of entering the West and the Canadian Northern equally desirous of entering the East, what could be more suitable than for the Canadian government to induce the two companies to establish a formal community of interest? Efforts were made to effect an arrangement on terms which to later generations have appeared most reasonable, but the attempt failed. In 1902, while efforts looking toward amalgamation had not yet been abandoned, the government granted permission (by 2 Edw. VII, c. 50) to the Canadian Northern to build

[27]See Glazebrook, *A History of Transportation in Canada*, pp. 315 ff.

eastward from Port Arthur to Montreal, Ottawa, and Quebec. In 1903 the government enacted its new policy concerning the Grand Trunk: a new company was to be incorporated, the Grand Trunk Pacific, a subsidiary of the Grand Trunk Railway. This Company was to build a line from Winnipeg via the Yellowhead Pass to the Pacific Coast with assistance by way of bond guarantees. The government was to build from Moncton, New Brunswick, by way of Quebec and through the east–west axis of the northern Ontario clay belt to Winnipeg. On completion, this section, the National Transcontinental railway, was to be leased to the Grand Trunk Pacific Company. It was completed late in 1913 at a cost which exceeded the original estimates by approximately $100 million, or by 200 per cent. The Grand Trunk Pacific declined to take up the lease on the National Transcontinental, and by September, 1914, when its own section was completed from Winnipeg to a newly created port on the Pacific, Prince Rupert, it was in a condition of acute financial embarrassment. In September, 1915, the last spike was driven in the Canadian Northern transcontinental line from Quebec to Vancouver.

By the early years of the First World War, therefore, there were three transcontinental railway lines in Canada. Of the three, only one, the Canadian Pacific, was in a financial position to survive the disruption of the war years without extensive assistance from the Dominion government. State assistance was necessarily so great that even the fiction of independent private enterprise in the Grand Trunk and the Canadian Northern systems came to appear untenable. The Canadian National Railway system was eventually established to operate, on government account, the Intercolonial, the Grand Trunk, the National Transcontinental, the Grand Trunk Pacific, and the Canadian Northern as a single railway system.

Since the various segments of the national policy were formulated and brought to maturity concurrently and interdependently it is difficult to determine the best sequence of analysis. Canadian tariff and transportation policies have always been closely correlated and this would suggest the advisability of considering tariff policy next, but protective tariffs were not introduced in Canada until more than a decade after Confederation. Meanwhile federal immigration and land policies had become firmly established. Although no purpose would be served by analysing Canadian immigration policy in the present study in detail, a few general points must nevertheless be made and elaborated briefly concerning it.[28]

[28]For a fuller discussion of this topic see V. C. Fowke, *Canadian Agricultural Policy* (Toronto, 1946), chaps. VI, VII.

Not all economic expansion in Canadian history had favoured immigration and settlement. Economic expansion as envisaged by the fur trader had no place for an active immigration policy. Immigrants merely intensified the problems of one-sidedness in the Atlantic shipping services, for the heavy cargo in the fur trade was westward, from Europe to the trading posts in the new world. Once established as agricultural settlers, the immigrants were a threat to the production and monopoly marketing of furs. The expansion of the timber trade in the nineteenth century, on the other hand, was quite compatible with immigration in so far as westward migration constituted a partial solution of the return-cargo problem in the Atlantic area: the heavy cargo moved eastward, from North America to Europe, and settlers along with settlers' effects served as a paying cargo for the return or westward voyage. Once settled, however, the immigrants were far from an unmixed blessing to the timber producer. Under careful control and in limited numbers they were useful as a source of winter labour and teams for the timber camps, and of feed for horses and provisions for men. But they were a constant threat to timber limits because the fires they used in the process of clearing their land often escaped control.

The economic purposes of the national policy were essentially commercial, and, to that extent, involved merely a continuation of the type of activities characteristic of the fur trade and the timber trade. But the commerce contemplated in the new policy was not only tolerant of but primarily dependent upon immigration and agricultural settlement. The national project would have failed without a vigorous and successful immigration policy. The foundations of Dominion immigration policy had been soundly laid by the province of Canada in the decade and a half preceding Confederation. The provincial Department of Agriculture became the Dominion Department of Agriculture at the time of Confederation and continued the promotion of immigration as its principal activity until 1892, when the Immigration Branch was transferred to the Department of the Interior. In addition to maintaining immigration agencies in Europe and preparing and distributing literature, the Dominion government introduced a system of passenger warrants and the payment of commissions to passenger agents. Despite the persistence which characterized the encouragement of immigration by the Dominion government, the results were far from encouraging until after the turn of the century.

The Liberal administration which was returned to power in the Dominion in 1896 intensified the promotion of immigration and of settlement in the Canadian West. Hon. Sir Clifford Sifton brought imagination and vigour to the administration of the Department of the Interior.[29] European immigration services were expanded and special efforts were made to attract settlers from the American Middle West. Unprecedented success accompanied these efforts throughout the decade and a half preceding the First World War. The tide of Atlantic and of North American migration turned toward Canada as it had not done since the middle of the nineteenth century.

The immigration policies of the Dominion government were obviously inseparably linked with settlement prospects, and these, in turn, with the availability of agricultural lands and with the Dominion's land-settlement policies. As already pointed out, the primary economic objective of the national policy was the establishment of a new frontier, an area where commercial and financial activity could readily expand and where labour and capital might find profitable employment. In terms of the economic competition of the day the requirement was for an agricultural frontier which could attract an adequate proportion of the annual flow of emigrants from the British Isles and Europe. The land which remained unoccupied in the Maritimes and in the province of Canada by the middle 1850's was unattractive in comparison to that in the Middle Western states. Of the $2\frac{1}{2}$ million emigrants who left European countries during the years 1853 to 1870, 61 per cent went to the United States, 18 per cent went to the Australian colonies, others went to Brazil and the Argentine, and a mere trickle arrived in Canada.[30] After the passage of the United States homestead law in 1862 it was clear that no land policy less generous would serve to divert the flow of European emigrants to a Canadian frontier.

The transfer of Rupert's Land in 1870 gave to the Dominion government a Canadian counterpart of the American public domain and the raw materials for the creation of a new agricultural frontier, but a new problem arose when a part of this newly acquired area was accorded provincial status as the province of Manitoba. According to the British North America Act (s. 92, ss. 5) the provinces were to retain possession and control over their unalienated lands. In order to maintain Dominion control over the western public domain, whether within provincial boundaries or in the territories beyond, the Dominion government found it expedient to introduce an inconsistency in federal land-control legislation. The Manitoba Act of 1870 provided that "All ungranted or waste lands in the Province [of Manitoba] shall be . . . vested in the Crown, and administered by the Government of Canada for the purposes of the

[29]See John W. Dafoe, *Clifford Sifton in Relation to His Times* (Toronto, 1931).
[30]Fowke, *Canadian Agricultural Policy*, p. 169.

Dominion." In the territories beyond the province of Manitoba the question of jurisdiction did not arise until the provinces of Saskatchewan and Alberta were created in 1905. The Manitoba principle was then extended to those provinces and the natural resources in the three prairie provinces remained under Dominion control until 1930. While, therefore, the four original provinces, and British Columbia and Prince Edward Island on entry into Confederation, all retained their natural resources without question at any time, the three prairie provinces acquired control over their public lands, and their natural resources in general, only after 1930.

The public lands of Manitoba and the Northwest Territories—or, after 1905, of the provinces of Manitoba, Saskatchewan, and Alberta—had to serve two purposes which could not be wholly harmonized. Under free homestead privileges these lands offered an attraction for immigrants and other settlers. But the Pacific railway had to be built in order to develop the West and to make possible the creation of a continent-wide nation. The Dominion government, in conformity with American example, gave assurance that the railway would cost the taxpayer nothing, that it would be built out of the revenue of land resources. The contractual grant of 25 million acres of the public domain to the Canadian Pacific Railway syndicate in 1881 marked the introduction of the railway land-grant system into Canada. The system was terminated in 1894 in the sense that no railway land-grants were made thereafter. The "earning" of land grants by railway construction, however, continued for years, and the selection and patenting of earned acreage was not concluded in Canada until 1908, by which date approximately 32 million acres of public lands had been alienated in western Canada as grants for railway construction.

Lands in Manitoba and the Northwest Territories had necessarily to finance railway construction as well as to attract immigrants. The proviso in all railway land-grants that the railway need select only those lands that were "fairly fit for settlement" meant that homestead entries were long delayed in certain areas pending the selection of such lands by the railways and, in the end, were often made available in areas in which the railways declined to accept any lands because of the opinion of their officers that such lands were not fairly fit for settlement.

It was originally proposed to survey the western territories into blocks containing 800 acres made up of four 200-acre lots. It was very early decided, however, to duplicate in Canada the American system of western survey because it was already familiar to potential immigrants by American example. This involved the survey and division of agricultural lands into "sections" one mile square, containing 640 acres, and the sub-division of these units into quarter-sections, one-half mile square, containing 160 acres each. Road "allowances," four rods in width, were provided one mile apart running north and south and ordinarily two miles apart running east and west. Thus every quarter-section—regarded as the basic unit of settlement—fronted on at least one road allowance. The sections were in turn combined into larger units, townships, six miles square (neglecting road allowances), and thus containing 36 sections of land. The townships, lying in north–south "ranges," were numbered consecutively northward from Township 1 in each range at the international boundary. The ranges were numbered consecutively westward from meridian lines.

Homestead regulations were promulgated in Dominion Orders in Council in 1871 and were incorporated in the Dominion Lands Act of 1872. Settlers were thus entitled to secure entry to a quarter-section of land on payment of a ten-dollar fee. Pre-emption rights were added in legislation in 1872. Under this legislation a settler, after homestead entry, could obtain an interim entry for an adjoining quarter-section and could purchase it at a price set by the government as soon as he had secured the patent for his homestead quarter. The privilege of pre-emption was withdrawn in 1894 after a clear and convincing demonstration that under the circumstances of the period it had contributed to speculation rather than to bona fide settlement. The privilege was restored in 1908.

The original regulations provided that homesteading was not to be permitted within the railway zone, a belt extending twenty miles on either side of the Pacific railway that was to be built. These regulations were changed from time to time and most significantly in 1879 and 1882 when, in effect, the even-numbered sections of all Dominion lands were opened for homestead entry. In areas where the railways selected their grants as odd-numbered sections, it developed, therefore, that homesteaders had railway lands adjacent to their individual holdings available for purchase in order to enlarge their farms beyond the original quarter-section size, or beyond the original half-section size in cases where the homesteader had pre-empted an additional quarter. Information which is available suggests that the Canadian Pacific Railway Company sold its lands over the years at prices which were better designed to encourage occupation and development of the West than to extract the maximum immediate return.

Homestead and railway lands did not make up the total area of the western provinces suitable for private settlement. Land was allocated in specific parcels throughout the entire area for the Hudson's Bay Company and for schools. As a part of the arrangement whereby the charter rights of the Hudson's Bay Company were relinquished, the Company was allowed to retain one-twentieth of the land in the "fertile belt," the latter defined in this case as comprising the total area between Lake Winnipeg and the Rockies and between the United States boundary and the North Saskatchewan River. This grant amounted to approximately 6.6 million acres. A further allocation of one-eighteenth of all western lands was made for educational purposes.

The policies of the Dominion government regarding land settlement, immigration, and railways had passed their formative stages, but little more, by 1900. Thirty years of effort had been required to lay firm foundations for the western wheat economy which was created within the first three decades of the twentieth century. Railway construction in the West before 1900 provided a strong stimulus to agricultural expansion and at the same time marked out the limits within which such expansion might take place: the completed railway was indispensable to the production and marketing of a bulky cash crop such as wheat, so settlement never advanced far beyond the actual or proposed "end of steel"; and the work of construction provided a source of cash income for settlers and potential settlers.[31]

While a substantial proportion of the settlement in the West before 1900 was effected by individual homesteaders, a much more significant feature was the establishment of colonies. A number of groups came unsolicited. In 1872, for example, a group of German Mennonites from the province of Berdiansk, Russia, investigated the possibilities of transfer to western Canada. They were granted financial assistance toward their passage, and settlements were reserved for them in the Red River Valley. By 1876 there were 6,150 of them in Manitoba. In 1874 the first reservations were made in western Canada for French Canadians repatriated from the New England states. In the ten-year period, 1876–85, 4,800 of these migrated to the West to form ten new settlements and to add to a number of others. In 1875 and 1876 an Icelandic colony was formed on the west shore of Lake Winnipeg. A wide variety of European groups was scattered throughout parts of the

---

[31]The building of canals and railways stimulated immigration and settlement in the St. Lawrence area before Confederation. The construction of the Canadian Pacific Railway's main line in the 1880's encouraged a wave of immigration but did not lead to commensurate settlement because the interval during which employment was available was too short. Brief references to the relationship between public works and settlement have been made in previous chapters, particularly chapter II.

plains area to form the nuclei for units of later expansion. On reservations or "nominal reservations," the latter without exclusive rights of entry to allocated territory, there were groups from England and Scotland and colonies of Hungarians, Scandinavians, Germans, Roumanians, Icelanders, Mennonites, Danes, Finns, Russians, Ukrainians, Belgians, and Jews. Many attempts at company colonization were made but, on the whole, proved unsuccessful.

Settlement did not spread uniformly before 1900. Southern Manitoba was occupied to the west of the Red River to include almost the whole of the area constituting the agricultural portion of Manitoba at the present day. From there, settlers spread into the southeastern part of the Northwest Territories and by 1900 had occupied a substantial part of the east and southeastern sections of the area which now comprises the province of Saskatchewan. Apart from this more or less consolidated region, settlement was concentrated in localities scattered from Manitoba to the foothills of the Rocky Mountains, in a ribbon along the line of the Canadian Pacific Railway, a wider strip from Edmonton southward to the boundary, and a cluster around the forks of the North and South Saskatchewan rivers. It was already noticeable that settlers tended to avoid the short-grass plains and to occupy its margins in preference to forming a pattern of continuous settlement from east to west.

By 1900 there were approximately 419 thousand persons in Manitoba and the territories. Of these, approximately 255 thousand were in Manitoba and 91 thousand and 73 thousand, respectively, were in the territories which today comprise the provinces of Saskatchewan and Alberta.

The final important feature of the national policy to be discussed here is the system of protective tariffs. The necessity for this feature, as for the others, was only imperfectly foreseen at the time of Confederation. A good deal was said about tariffs in the *Confederation Debates* but the discussion had to do with the advantages which were anticipated from the removal of tariffs between the provinces and the resultant establishment of a free-trade area containing some millions of people. Federation removed the possibility of imposing tariffs along interprovincial boundaries but created *international* boundaries for the combined British North American territories and raised the question of trade policy for the Dominion as a whole. The Maritimes in the pre-Confederation years believed more firmly in low tariffs than did the Anglo-Saxon merchants in the United Province of Canada. Their tariffs were lower, and they had a strong bargaining position in Confederation politics in that they had less to lose by staying as they were and less hope of what might be gained by the change. The original tariff rates for the new Dominion of Canada vis-à-vis the world were established by lowering the rates in the province of Canada rather than by raising those in the Maritimes.

Railroads and tariffs are more clearly interrelated in the national policy than are any of the other elements. It is clear that Canadian rail lines linking central Canada with the Maritimes and the West provided only the physical facilities for the movement of goods: they did not in themselves make it certain that manufacturing facilities would develop in the central provinces to supply the outlying regions. Without protective measures of some sort, manufacturers in central Canada could secure and hold the markets of the outlying regions only if they could deliver goods in competition with the highly efficient mass-production industries of the eastern and middle-western United States. This they have consistently alleged they could not do. Whether they could or not has never been put to the test. A policy of tariff protection was instituted in Canada before there were any significant outlying markets to supply, and, indeed, before there was any great body of industry to supply them, and this policy has been maintained without basic modification to the present day.

The protective element of the national policy was enacted in 1879, but the tariff policy of the province of Canada before Confederation foreshadowed it. In the Cayley-Galt tariffs of 1858 and 1859 the import rates of duty on manufactured goods were considerably increased. In the budget of 1859 the customs rate on the general list of merchandise was placed at 20 per cent ad valorem. While an increase in the rates on manufactured goods was clearly compatible with the *letter* of the reciprocity agreement with the United States, it was nevertheless regarded in American circles as a breach of its spirit.[32] This view had lasting adverse significance for Canadian-American tariff negotiations.

Galt justified the tariff increases on the basis of the costs and benefits of improvements in transportation undertaken by the government. His reasoning provided a clear and early statement of the principle of the interdependence of tariffs and transportation facilities in Canada. The

[32]It appeared to the Americans that the increases in rates of duty were made applicable to a particularly large number of products normally imported from the United States. Also certain rates of duty were changed from specific to ad valorem and products were to be dutiable on the basis of their value in the market where they were last bought instead of in the country of origin. The latter provision was deliberately designed to favour direct importation of such products as tea, coffee, molasses, and sugar via the St. Lawrence instead of indirectly via the United States. See Donald C. Masters, *The Reciprocity Treaty of 1854* (London, 1936), p. 113 ff.

construction of canals and railways in new countries such as Canada, he argued,[33] was extremely costly to the government but was, at the same time, of great benefit to producers and consumers. Such improvements raised the net returns of export producers and lowered the cost of imports to consumers. The imposition of tariffs, therefore, was but a reasonable method by means of which the government might recoup a part of its developmental outlay. Galt described the policy represented by the 1859 changes as one of "incidental protection," a policy of tariffs primarily for revenue but "affording at the same time an incidental amount of protection" to domestic manufacturers.

If the provincial tariff of 1859 was designed primarily for revenue, and incidentally for protection, the Dominion tariffs enacted in 1879 and subsequent years were designed primarily for protection, and incidentally for revenue. This change in emphasis was essential to the national purpose. Construction of a Pacific railway would make possible the economic development of the West. Protective tariffs would foster interprovincial trade in place of international trade. Canadian manufacturers would be assured as fully as possible of exclusive rights to the total Canadian market. Together, railways and tariffs would integrate the expanding area of economic activity. Tariffs would ease the burden of improvements in transportation by providing railway traffic and a more diversified economy as a source of tax revenues.

Canadian political leaders widely separated in time and in party affiliations have expressed the philosophy of creating a closely integrated economy by means of national policies. McGiverin stated the objective clearly during the *Confederation Debates* in 1865.[34] Although he did not specify railways or tariffs, or indeed any particular instrument for the development of the West, he nevertheless indicated with the greatest clarity the role assigned to the western territories at the time of Confederation.

Macdonald's interpretation of a national policy of protective tariffs was perhaps most fully outlined in the session of 1878 while he was still in opposition. On March 7 of that year, as the House moved to go into committee on the budget, he introduced the resolution that ". . . this House is of the opinion that the welfare of Canada requires the adoption of a National Policy, which, by a judicious readjustment of the Tariff, will benefit and foster the agricultural, the mining, the manufacturing and other interests of the Dominion; . . ."[35] His arguments in support of the resolution concerned the development of resources, the main-

[33]See Canada, *Sessional Papers*, 1862, no. 23.
[34]Pp. 469–70.
[35]Canada, *House of Commons Debates*, 1878, p. 854.

tenance of opportunities for employment, the prevention of Canada's continuing as a "slaughter market" for American merchandise and, finally, the stimulation of interprovincial trade.

> The resolution speaks not only of a reasonable adjustment of the tariff but of the encouragement and development of interprovincial trade. That is one of the great objects we should seek to attain. Formerly, we were a number of Provinces which had very little trade with each other, and very little connection. . . . I believe that, by a fair readjustment of the tariff, we can increase the various industries which we can interchange one with another, and make this union a union in interest, a union in trade, and a union in feeling. We shall then grow up [sic] rapidly a good, steady and mature trade between the Provinces, rendering us independent of foreign trade, and not as New Brunswick and Nova Scotia formerly did, look to the United States or to England for trade, but look to Ontario and Quebec,—sending their products west, and receiving the products of Quebec and Ontario in exchange. Thus the great policy, the National Policy, which we on this side are advocating, would be attained.[36]

Macdonald made the introduction of protective tariffs a major issue in the 1878 election campaign. Asked at Hamilton what protection he proposed to give he said, "I cannot tell what protection you require. But let each manufacturer tell us what he wants, and we will try to give to him what he needs."[37] The Conservative *Mail* of Toronto urged the electors in 1878 to get rid of the "starvationists," the Liberals, and to "bring back a rich prosperity to Toronto and the Dominion at large."[38]

In 1879, Sir Leonard Tilley as Minister of Finance introduced the tariff rates of the national policy in the first budget of the newly elected Parliament. In summing up the immense detail of the tariff changes he said:

> . . . it does appear to me, Sir, that . . . the time has arrived when we are to decide whether we will be simply hewers of wood and drawers of water; whether we will be simply agriculturists raising wheat, and lumbermen producing more lumber than we can use, or [that] Great Britain and the United States will take from us at remunerative prices; whether we will confine our attention to the fisheries and certain other small industries, . . . or whether we will inaugurate a policy that will, by its provisions, say to the industries of the country, we will give you sufficient protection; we will give you a market for what you can produce; we will say that while our neighbors build up a Chinese wall, we will impose a reasonable duty on their products coming into this country; at all events, we will maintain for our agricultural and other productions largely, the market of our own Dominion.[39]

36*Ibid.*, p. 861.
37As cited in Edward Porritt, *Sixty Years of Protection in Canada, 1846–1912* (Winnipeg, 1913), p. 260.
38*Ibid.*, p. 261.
39Canada, *House of Commons Debates*, 1879, p. 429.

During the first two decades under the new protective tariff, the great bulk of Canadian activity was confined to the central and maritime provinces. In framing the protective schedules of 1879 the goal had been, in Tilley's words, to find "the best means of reducing the volume of our imports from all parts of the world." Subsequent modifications in the original national policy rates were made in furtherance of this aim. By the turn of the century the scale of economic activity and the expansion of market prospects in the prairie provinces were of such magnitude that the exclusion of foreign imports from that area took on special significance. For this purpose the tariff structure of the national policy required only to be maintained. This the Liberals, under Sir Wilfrid Laurier, found it expedient to do despite pledges made during the campaigns of the 1890's.

Sir Wilfrid himself has left us one of the clearest statements on record of the place occupied by the prairie provinces in the framework of protection built by the national policy:

They [the settlers in western Canada] will require clothes, they will require furniture, they will require implements, they will require shoes—and I hope you can furnish them to them in Quebec—they will require everything that man has to be supplied with. It is your ambition, it is my ambition also, that this scientific tariff of ours will make it possible that every shoe that has to be worn in those prairies shall be a Canadian shoe; that every yard of cloth that can be marketed there shall be a yard of cloth produced in Canada; and so on and so on. . . .[40]

It is not possible to measure the over-all effect of tariff policy on the growth and integration of the Canadian economy, nor would it be relevant to the present purposes. The tariff is inseparably tied up with transportation policy in the development of the Dominion and the significance of the tariff varies as between regions in Canada. No attempt will be made to measure this variation but its existence and nature can be clearly established.

In explaining the regional incidence of tariffs under the national policy, it is important to recall, first, that they were introduced before there was any considerable economic development in the prairies, and second, that they remained relatively unchanged until as late as 1930. The effect of the imposition of a new duty or an increase in an old one is different from the effect of the continuance of an old one. An economy which develops within a framework of tariffs already established will differ from one which develops under circumstances similar except for

[40]From a speech made to the Canadian Manufacturers' Association, Quebec City, 1905, quoted in *Canadian Annual Review of Public Affairs*, 1905, pp. 149–50.

the absence of such a framework. The development will, from its beginning, take cognizance of the effects of the tariff, as nearly as such effects can be foreseen. Types and ranges of activity rendered uneconomic by the tariff will never start, except in error. If the same tariff were to be imposed on an economy already well developed, the resultant readjustment and alteration of equities might well be drastic in degree.

The prairie economy grew up within a pre-established framework of tariffs which shaped, limited, and curtailed its development. With the possible exception of the Red River Valley, however, where settlement was well established before 1880, it cannot be said that the introduction of protective tariffs destroyed any equities already created by economic activity in the prairie region. Nevertheless a differential regional incidence of the Canadian tariff can readily be demonstrated.

Duties on imports into Canada have curtailed a wide range of industrial importations, particularly from the United States, and have replaced them with higher-priced Canadian products. To the extent that this has happened, that is, to the extent that the national policy of protection has been successful, Canadian industry has expanded to a greater extent than would otherwise have been the case. At the same time, costs of production and of living have been increased to a degree represented by a variable proportion of the amount of tariff protection. Since, for geographical reasons, the greatest industrial opportunities are concentrated in the central provinces, the expansion in industrial activity attributable to tariff protection has taken place in the central provinces. Under the conditions which have prevailed in the past the prairie provinces have had few industrial possibilities but great capacity for export production. The tariff, then, has provided no scope for western industrial expansion and has had the unmitigated effect of curtailing the expansion of export activity because of the pronounced increase in costs to which it gives rise.

The differential regional effects of protective tariffs in Canada are not to be sought in long-run inequalities in the returns to labour or capital, or in profit margins. To the extent that Canadian industrialists require protection against foreign competition, that protection is eaten up in inefficiency and high costs rather than in high profits. If the full degree of protection that exists is not needed, domestic producers will nevertheless secure abnormal profits only if they exploit their protected position with the aid of effective monopoly.

The differential regional effects of protective tariffs are to be found in property values. By fostering industry, Canadian tariffs have increased the value of properties best suited for industrial plants and for the housing of industrial labourers. By raising the costs of production and living in the export areas, Canadian tariffs have restricted the expansion of these areas and have to that extent limited the relative increase in their property values. This is of particular significance in the prairie provinces where production for export is of such overwhelming importance and where property equities are so widely diffused among the resident population.

The fact that the tariff has operated to provide a financial subsidy to the population of the areas of Canada in which it has stimulated industrial expansion, notably in the central provinces, is in no way incompatible with the likelihood that the subsidy has been unequally apportioned among the various segments of the population within these provinces. Wage earners as such have suffered rather than gained because of the tariff, but property owners have achieved permanent and substantial gains. Taxing bodies at both the local and provincial levels in the industrial region have secured an enlargement of their bases of assessment on property and business as a direct result.

The industrialization of central Canada is not, of course, solely due to the protective tariff. The central provinces have great natural advantages over other parts of the Dominion for industrial activities and for certain types of development they alone possess the necessary conditions. Nevertheless, regional specialization rests on acquired as well as natural advantages. Of two regions, the one with a prior start in industrialization may, because of that fact alone, have a permanent superiority over the other. For decades after the national policy of protection began to foster manufacturing in the central provinces the prairie provinces were without industry to foster. The lack of industrialization in the West today is due in part to the impossibility of achieving competitive competence as against eastern industries long since firmly established with the aid of protection.

When representatives of the prairie provinces point out the unequal regional impact of the tariff structure it is commonly argued in reply that tariffs and railways go together in the national policy and that the effects of the one cannot be analysed apart from the other. Western development of any extent would obviously be impossible without railway facilities and, so it is argued, the tariff structure assures the east–west movement of goods which, in turn, serves as the paying traffic for the Canadian railway system.

Without the tariff the Canadian system of railways would be wholly uneconomical. It would be incorrect to assume, however, that the prairie provinces would be without adequate railway facilities had the Canadian transcontinentals and their feeder systems not been built. One of the chief concerns of the early railway policy of the Dominion government was the exclusion of American railways from Canadian territory to the west of the Great Lakes. The management of the Northern Pacific and Great Northern railways stubbornly persisted, from the 1860's till the end of the century, in their attempts to build into the Canadian territory both on the plains and in the mountains. Their aim was to draw the whole western Canadian region into their commercial and general economic orbit. The national policy of tariffs and railways was successful in preventing this absorption. As far as the western provinces are concerned, therefore, Canadian railways are expensive alternatives to American railways rather than to no railways at all.

CHAPTER FIVE

Establishment and Growth of the Wheat Economy

ALTHOUGH prerequisite to western development, the national policy was not by itself sufficient to make such development possible on the scale eventually attained. The establishment of the prairie wheat economy, which may be regarded as its first major economic triumph, was accompanied by tremendous expansion throughout the entire Canadian economy and was an integral part of a complex of dynamic forces which pervaded the western world. Professor Mackintosh has spoken of the "conjuncture of favourable circumstances"[1] which marked the transition from the nineteenth to the twentieth century and which gave to Canada three decades of unprecedented expansion. This conjuncture of world circumstances created the opportunity for Canadian expansion, but a half-century of foundation work along the lines of the national policy had prepared Canada for the opportunity.

Technological changes associated notably with iron and steel and power spread throughout the nations contiguous to the North Atlantic in the latter part of the nineteenth century with ramifications sufficient to merit description as a new industrial revolution. Industrialization and urbanization proceeded apace. The accompanying growth and concentration of population called for the assembly of raw materials and overseas food supplies. Improvements in iron and steel technology and in mass production methods contributed to a relative decline in the prices of manufactured goods and transportation services. Ocean freight rates at the end of the century were approximately half what they had been thirty years before and continued to decline into the twentieth century. Interest rates in 1897 were the lowest in recorded history,[2] signifying the existence of ample supplies of mobile capital in search of profitable areas of exploitation. Cumulative increases in the world supply of gold associated with technological and geographical discoveries contributed to

---

[1] W. A. Mackintosh, *The Economic Background of Dominion-Provincial Relations* (Ottawa: King's Printer, 1939), p. 24 and *passim*; see also *Report of the Royal Commission on Dominion-Provincial Relations* (Ottawa: King's Printer, 1940), Book I, pp. 66 ff.

[2] *Report of the Royal Commission on Dominion-Provincial Relations*, p. 67.

advances in prices which altered cost-price relationships in favour of the countries that produced raw material and raw food. The relative inferiority of unalienated lands in the United States placed a progressively mounting premium on those available for homestead or purchase in the prairie provinces of Canada.

The incidents associated with the establishment of the Canadian wheat economy have been fully recounted in other studies[3] and there is no need to repeat them in detail here. It is necessary, however, to review the main economic processes associated with the creation of the Canadian West in order to emphasize their dynamic nature and to indicate their impact upon Canadian economic life. The prospect for the profitable employment of capital and labour in the production of wheat on the Canadian plains attracted millions of immigrants to the continent and to the region, and prompted the investment of billions of dollars not only in the prairie provinces but throughout the entire nation. The prairie provinces constituted the geographic locus of the Canadian investment frontier in the first three decades of the twentieth century. The dynamic influence of the frontier permeated and vitalized the Canadian economy and extended far beyond. Pertinent here is the reminder that ". . . the economic frontier of any country must always be conceived of not in terms of its own boundaries, but in terms of the possibility of capital investment throughout the entire world."[4]

An investment frontier may be geographically diffused but it nevertheless has tangible, concrete expression in the process of real-capital formation. The establishment of the wheat economy required the assembly in the prairie provinces of a massive structure of capital equipment without which the large-scale production and marketing of wheat would have been impossible. This included not only the equipment of the farms but also the equally indispensable equipment of the market centres throughout the region and of the transportation routes between. Each one of the hundreds of thousands of new farm units in the prairie provinces had to be provided with buildings: a house or some sort of dwelling, however rudimentary, a barn or stable, and granaries; a complete if variable set of farm machinery for cultivating, seeding, harvesting, and threshing; and, in addition, a wide variety of incidental capital such as

[3]See: W. A. Mackintosh and W. L. G. Joerg, eds., Canadian Frontiers of Settlement Series (Toronto: The Macmillan Company of Canada) of which eight out of nine proposed volumes were published; Report of the Royal Commission on Dominion-Provincial Relations, Book I; W. A. Mackintosh, The Economic Background of Dominion-Provincial Relations; Report of the Royal Commission on the South Saskatchewan River Project (Ottawa: Queen's Printer, 1952), part II, chap. III, "The Historical Setting."
[4]A. H. Hansen, Fiscal Policy and Business Cycles (New York, 1941), p. 44.

household furnishings, fencing materials, pumps, windmills, and hand tools. The marketing of farm produce and the purchase of equipment and supplies by the farming population required thousands of miles of additional railway structure and hundreds of new market centres, each of the latter equipped with sidings, elevators, and loading platforms, warehouses and stores, and housing for the local residents.

The capital equipment for the prairies came largely from other parts of Canada. Tariff policy—as it was specifically designed to do—diverted to the provinces of Ontario and Quebec much of the demand for machinery, tools, hardware, articles of leather, clothing, and home furnishings which would otherwise have been supplied by American manufacturers. Buildings and structures of all kinds were, of course, assembled on the spot but the lumber came from outside the prairie region, for the most part from British Columbia.[5] All parts of the Dominion with the exception of the maritime provinces expanded their industrial and other economic activity in direct response to the opening of the prairie market. Railways moved equipment and building materials to the prairies and transported grain to eastern and export terminal markets.

The movements and growth of population provide a significant index of the magnitude of the opportunities for investment. Within thirty years after 1900 the population of Canada all but doubled, to total 10,377,000 in 1931. Meanwhile the population of the prairie provinces increased more than fivefold, to a total of 2,354,000. In 1901 only 8 per cent of

TABLE I

POPULATION OF THE PRAIRIE PROVINCES AND PERCENTAGE OF RURAL TO TOTAL POPULATION, 1901–51

	Man.	Sask.	Alta.	Total	Rural Population as % of Total
1901	255,211	91,279	73,022	419,512	75
1906	365,688	257,763	185,195	808,646	70
1911	461,394	492,432	374,295	1,328,121	65
1916	553,860	647,835	496,442	1,698,137	64
1921	610,118	757,510	588,454	1,956,082	64
1926	639,056	820,738	607,599	2,067,393	64
1931	700,139	921,785	731,605	2,353,529	62
1936	711,216	931,547	772,782	2,415,545	63
1941	729,744	895,992	796,169	2,421,905	62
1946	726,923	832,688	803,330	2,362,941	61
1951	776,541	831,728	939,501	2,547,770	51

SOURCE: Dominion Bureau of Statistics, *Census Reports.*

[5]Prior to 1914, 70 per cent of the output of lumber in British Columbia went to the Canadian prairie market. The proportion was not as high in the inter-war years. *Report of the Royal Commission on Dominion-Provincial Relations,* Book I, p. 123 n.

all Canadians lived on the central plains. By 1931 nearly 25 per cent were there. Table I indicates the extent and rapidity of population growth in the prairie provinces after the turn of the century and shows clearly the relative stability of the total in the twenty-year period after 1931.

Homesteading activity in western Canada was particularly concentrated in the years between 1900 and the outbreak of the First World War. Approximately 675,000 homestead entries were made in the prairie provinces from the early 1870's, when the principle of homesteading was introduced into Canada, until 1930, when the Crown lands in the prairie provinces were transferred to provincial jurisdiction. Over 440,000 of these entries were made in the fifteen-year period preceding the First World War. In the five-year interval, 1900 to 1904 inclusive, more entries were made than in the preceding quarter of a century, and even by 1905 Canadian homesteading had not begun to approach its maximum rate. The four years from 1909 to 1912 witnessed the greatest sustained establishment of new homesteads, with an average of over 40,000 a year. During these years the short-grass plains between Moose Jaw and Calgary, the driest part of the prairie provinces, were thrown open for homesteading and pre-emption. The entire area was quickly occupied.

Table II sketches the main outlines of the agricultural occupation of the Canadian plains after 1900 in terms of numbers, size, and state of improvement of farm units. The rate of settlement and land improvement was particularly rapid from 1901 to 1911 but remained high until

TABLE II

NUMBER AND AREA OF FARMS AND ACREAGE UNDER FIELD CROPS IN THE PRAIRIE
PROVINCES, 1901-51

	Number of farms (thousands)	Area of occupied farms (millions of acres)	Average size of farms			Area of improved land (millions) of acres)	Area under field crops (millions of acres)
			Man. (acres)	Sask. (acres)	Alta. (acres)		
1901	55.2	15.4	274.2	285.1	288.6	5.6	3.6
1906*	—	—	—	—	—	—	—
1911	199.2	57.7	279.3	295.7	286.5	23.0	17.7
1916	218.6	73.3	288.5	353.8	339.3	34.3	24.6
1921	255.6	87.9	274.5	368.5	353.1	44.9	32.2
1926	248.2	88.9	270.6	390.1	370.5	49.3	35.0
1931	288.1	109.8	279.2	407.9	400.1	59.8	40.0
1936	300.5	113.1	271.4	399.6	403.9	60.9	40.2
1941	296.5	120.2	291.1	432.3	433.9	65.5	38.4
1946	269.6	117.6	306.2	473.0	462.9	65.4	41.7
1951	248.7†	123.9	338.5	550.5	527.3	71.8	43.4

*Not available.
†Not directly comparable with previous data due to change of definition.
SOURCE: Dominion Bureau of Statistics, *Census Reports.*

1931. Expansion did not come to an end suddenly in 1931, but a consideration of the various data presented by Table II suggests a contraction in rates of growth at that time which has not been restored to the present day. The number of farms in the prairie provinces reached a maximum in 1936 and consolidation of holdings by lease and purchase into larger units has reduced the number appreciably since that time. The continuous increase in the average size of farms is particularly noticeable in the data for Saskatchewan and Alberta. The period from 1921 to 1926 merits comment as the first to show widespread retrenchment in the prairie wheat economy. During this interval the number of farms declined and the area of occupied farms and of improved acreage increased but slightly. Throughout the same years many farms were abandoned, particularly in eastern Alberta and southwestern Saskatchewan.

Settlers on prairie farms used the cultivated acreage almost wholly for wheat as the cash crop and for coarse grains to be fed to livestock on the farms. Tractors were of relatively little importance until the inter-war years, and, as horses were necessarily used for power, a part of the farm acreage had to be utilized in the production of feed. Beef and dairy cattle and hogs were of considerable importance in certain areas, but for the most part did not appreciably affect the allocation of cultivated acreage.

Increased wheat acreage provided the real basis for prairie agricultural and economic expansion. Plantings (practically all in the prairie provinces) increased from 4.3 million acres in 1901 to 11.1 million in 1911, 23.3 million in 1921, and 26.4 million in 1931. Shipments abroad increased from small amounts in the 1890's till by the end of the First World War they were the largest single Canadian export in terms of dollar value. Wheat and flour exports yielded $10.9 million in 1901, $377.5 million in 1921, and $495 million in 1929. Canada exported more than a million bushels of wheat a day on the average in 1921. Table III indicates the production trends. It will be seen that more than 90 per cent of the Canadian total is grown each year in the prairie provinces. It will also be noted that the wheat acreage of the prairie provinces has remained relatively stationary over the years since 1931, a figure of approximately 25 million acres having considerable claim to being the long-run normal planting.

The establishment of the wheat economy did not proceed uniformly throughout the first thirty years of the twentieth century. The development of the prairie region was influenced by marked irregularities in climatic and geographic circumstances, and the rate of economic advance upon the frontier was subject to the effects of the First World War and its destructive aftermath. The wheat growing area of the West is entirely

settlement correspond to this threefold division. Settlers first occupied the park belt with its familiar and seemingly indispensable complement of wood, water, and hay. By 1900 there was a clear indication of the tendency for settlement to swing northwestward in an arc out of the Red and Assiniboine valleys in the east, and northward along the margins of the Rocky Mountain foothills, avoiding the open plains between. The early years of the new century saw settlement moving inward from this arc, spreading out upon the open plains, but still avoiding the more nearly arid areas in the centre. The Canadian Pacific Railway Company declined to accept any significant portion of its land grant in the area from Moose Jaw to Calgary.[6] In 1908 the Dominion government finally terminated the railway land-grant process of selection and simultaneously reintroduced the pre-emption principle into land-grant policy. In 1909 the entire central zone, the most nearly arid area of the entire plains, was thrown open for homestead and pre-emption entry. The pre-emption area with which the new regulations were concerned was defined as the territory from Moose Jaw on the east to Calgary on the west, and from the international boundary on the south to Battleford on the north. The years after 1908, with the exception of 1910, were years of better than average rainfall, and before the outbreak of the First World War the last and driest portions of the southern part of the prairie provinces were finally occupied. Railway building added to the opportunity and incentive for settlement. Approximately 860 miles of railway were built south of the South Saskatchewan River in the five-year period after 1910.

The indiscriminate settlement of the dry belt—or dry triangle—which took place under the false stimulus of abnormally favourable moisture conditions after 1908 had serious regional consequences a decade later. Before noting the outcome, however, it is necessary to comment on the progress of the wheat economy as a whole following the outbreak of the First World War. The investment boom which had characterized the early years of the twentieth century had exhausted itself by 1913, and serious economic difficulties faced Canada and the prairie provinces as a result. Demands created by the war averted the worst of the dilemma, however, and opened for Canada a new era of expansion more pronounced even than that which had gone before. The demand of the Allies for munitions took up the slack in Canadian industry. The demand for foodstuffs called for further expansion rather than retrenchment in the wheat growing area and induced the cultivation of millions of acres which

[6]Except, of course, for the tract of 2,900 thousand acres which they accepted in a block between Medicine Hat and Calgary in 1903 and which they developed as an irrigation project. See chap. IV, n. 23.

had formerly been occupied but not improved. Between 1913 and 1919 the area devoted to wheat increased by 80 per cent and the acreage used in the production of field crops increased by an amount equal to the total increase of the preceding twenty years.[7]

The wartime boom was accompanied by an inflation of prices which was continued and intensified in the months succeeding the armistice of 1918. The peak of the boom came during the first half of 1920 and a sharp recession thereafter carried price indices drastically downward. The severity of the depressive forces displayed the usual inequalities of impact, with agricultural and raw-material producers in general suffering to a greater extent than industrial producers and processors. From 1920 to 1923 the prices of farm products fell by one-half and the price of wheat fell by almost three-fifths, whereas the prices of manufactured goods fell by but one-third and the Canadian cost of living index by less than one-fifth.[8] Under these circumstances the over-all economic position of Canadian wheat producers can readily be imagined. Heavily in debt for land and equipment, as is normal for frontier producers, prairie wheat growers were thrust within a matter of months during 1920 into the position of having their obligations doubled and even trebled, in real terms, by the collapse of agricultural prices. There was little comfort in the fact that prices were, at the worst, falling merely to their pre-war level. The wartime expansion in wheat production, the improvement of land, and the acquisition of the necessary capital equipment had taken place at greatly inflated price levels. Farm mortgages, and indebtedness to banks, machine companies, and local merchants were all distorted by measurement in terms of wartime dollars.

The calamitous economic pressures of the early 1920's were rendered more acute by prolonged drought. Climatic conditions had been irregular for a series of years after the fabulously bountiful crop of 1915. There was adequate moisture in 1916 but the quality of the crop was seriously impaired by rust. Moisture conditions were uneven in 1917, varying greatly from region to region. The next four years constituted a period of widespread drought. For substantial portions of the Canadian semi-arid plains there were five crop failures in a row, from 1917 to 1921 inclusive.

The impact of the drought was most severe in the dry core of the plains area. The census records of the growth of population and the abandonment of farms offer striking evidence of the enforced retrenchment in

[7]*Report of the Royal Commission on Dominion-Provincial Relations*, Book I, pp. 90–1.

[8]*Ibid.*, p. 113.

this region. A group of eleven census districts,[9] contiguous to the Saskatchewan boundary in Alberta and occupying a great block in southwestern Saskatchewan, had a population of approximately 395,000 in 1921 and slightly less than that in 1926. The heavy immigration of the ten or twelve years before 1921 had been reversed and converted into an exodus which, during the years from 1921 to 1926, was of a magnitude equal to the entire natural increase in the area. The retreat from the land was much more pronounced in Alberta than in Saskatchewan. Of the seven Saskatchewan census districts within the area, all except one (number 7, west of Moose Jaw) had at least a slight increase in population during the five-year period, 1921 to 1926, and as a group their population increased by 20,000. In Alberta, on the the other hand, all four of the census districts declined absolutely in population. Division 3, comprising a large area north of Medicine Hat, declined by 30 per cent.[10]

[9]Census districts 2, 3, 4, 7, 8, 12, and 13 in Saskatchewan, and 1, 3, 5, and 7 in Alberta. See *Report of the Royal Commission on the South Saskatchewan River Project*, pp. 106–7.

[10]The Tilley East area, covering approximately 1½ million acres, coincides roughly with the eastern half of census division no. 3 in Alberta. It was the first of the special areas established by the Alberta government to provide a settlement programme which might correct the errors of homesteading and pre-emption in the driest part of the entire Canadian plains region. Farm and village residents and public agencies within the Tilley East area were in such desperate financial circumstances by the middle twenties that the Alberta Legislature appointed a commission in 1926 to investigate, and recommend corrective measures. The commission reported that agricultural settlement had gradually displaced ranching in the area in the years immediately before and after 1910, that there had been few good crops except for those of 1915 and 1916, and that intolerable burdens of debt had been accumulated. The commission found that at the peak there had been approximately 2,400 resident farmers in the district but that the number was already reduced and should be still further reduced. They pointed out that "a considerable percentage of the land alienated from the Crown in the right of the Dominion of Canada [was] passing to the Crown in the right of the Province of Alberta through the failure of the owners . . . to pay taxes levied against their lands, and that the claims registered against such lands were far in excess of their value." The commission recommended a complete cessation of land alienation in the area, the disorganization of existing municipalities and inactive school districts, and a consolidation and reorganization of settlement under a joint Dominion and provincial board. The province effected the recommendations of the commission in 1927, by *Statutes of Alberta*, 1927, c. 45. The joint board operated from 1929 to 1931 when, following the transfer of natural resources to the province, the area came under the control of a provincial administrator in the Department of Municipal Affairs. By the early 1930's only approximately 500 farmers were in the area as compared with the peak total of 2,400 a decade or so before. Cf. G. A. Elliott, "Problems of a Retrograde Area in Alberta" in W. A. Mackintosh, *Economic Problems of the Prairie Provinces* (Toronto, 1935), Appendix B, pp. 291–4; also "Report of the Commission on the Tilley East Area," Province of Alberta, *Sessional Papers*, 1927, no. 20 (Edmonton: King's Printer).

The four divisions as a group had approximately 20,000 fewer residents in 1926 than in 1921. They contained almost 20 per cent of the total provincial population in 1921 but less than 17 per cent in 1926.

Loss of population from the area under consideration here is more specifically related to agricultural failure if we take note of the abandonment of farms with which it was associated. There were 4,900 abandoned farms in Saskatchewan in 1926 but they were not concentrated in any particular area except for a group of some 900 in census division number 8, next to the Alberta boundary. At the same time there were 118,000 operating farms in the province and the proportion of abandonment may not be regarded as exceptionally high. In Alberta, however, the situation was more acute. With 77,000 farms operating in 1926 there were 10,400 abandoned farm units. Over 7,300 of these were concentrated within the four census divisions along the Saskatchewan boundary and 5,000 of them were in divisions 3 and 5. Thirty per cent of the farm units in the four divisions were abandoned. In division number 3, over half the farm units and farm acreages were abandoned. The proportion was particularly high among the quarter-section units and heavy as well for the half and three-quarter section units.

The economic distress which characterized the early part of the 1920's formed the context within which major political and organizational developments occurred within the wheat economy. The major political features of the early inter-war years are dealt with in other volumes in this series.[11] It is only intended at this point to call attention to the general economic situation in which certain important policy decisions were made.

Moisture conditions gradually improved throughout the entire wheat growing region. Good crops in the years 1925 to 1927 were followed by the "bumper" crop of 1928. Over the same period, cost-price relationships were much more favourable to agricultural producers of all kinds than they had been in the first half of the decade.[12] The wheat economy prospered again. Immigration revived and agricultural settlement moved northward in the prairie provinces with a pronounced regional concentration in the Peace River area in Alberta. Along with the new out-

[11]W. L. Morton, *The Progressive Party in Canada* (Toronto, 1950); C. B. Macpherson, *Democracy in Alberta* (Toronto, 1953).

[12]"By 1926 the average of the prices of all exports, of wheat, newsprint, and lumber were 56, 70, 76, and 48 per cent respectively, above those of 1913, compared with increases of 37, 54, 45, and 48 per cent respectively, in the prices of all imports, manufactures, iron and steel and railway freight rates." *Report of the Royal Commission on Dominion-Provincial Relations*, Book I, p. 114 n. Between 1920 and 1924 the total cost of shipping wheat from Regina to Liverpool declined from 67 cents to 33 cents per bushel. *Ibid.*, p. 120.

thrusting of the agricultural frontier went the customary complement of railway building and the mushroom growth of market centres. Approximately 2,900 miles of railway line were added in the prairie provinces between 1920 and 1930, with practically no increase in the milage in any other part of Canada.[13]

The comparative prosperity of the later 1920's gave considerable impetus to the first major mechanical revolution to take place in the wheat economy.[14] Table IV provides indices of this revolution in terms of tractors, trucks, and harvester combines. The gasoline tractor was well established in the West by the end of the First World War but its use became much more general in the later twenties. The harvester combine was practically unknown on the Canadian plains until after the middle of the twenties. The farm motor truck first appeared in significant numbers in the wheat economy in the same interval.

TABLE IV

NUMBER OF TRACTORS, HARVESTER COMBINES, AND MOTOR TRUCKS IN THE PRAIRIE PROVINCES, 1921–51

	Tractors	Harvester combines	Motor trucks
1921	38,485	—	—
1926	50,136	—	5,640
1931	81,659	8,897	21,517
1936	81,657	9,827	21,293
1941	112,624	18,081	43,363
1946	151,161	38,870	56,177
1951	236,930	79,117	113,512

SOURCE: Dominion Bureau of Statistics, *Census Reports*.

The prosperity of Canadian wheat producers and the further expansion of the wheat-growing frontier during the later 1920's had much in common with the prosperity and expansion of earlier years. Contemporary observers may well have regarded these conditions as merely a continuation of the developments which had been interrupted by the war and the depression that followed it. Immigration, homestead settlement, railway construction, the establishment and growth of market centres,

[13]Sixteen hundred miles were added in Saskatchewan, 1,200 miles in Alberta, and 100 in Manitoba. Cf. *ibid.*, p. 121 n.

[14]In terms of the categories of Professor Hansen, this would constitute intensive investment in contrast to the extensive investment associated with the growth of population and the discovery of new resources. Professor Buckley calls attention to the shift in the relative importance of the two segments of the investment process in western Canada. The extensive aspect was more pronounced *before*, and the intensive aspect *after*, the period 1915–20. See K. A. H. Buckley, *Capital Formation in Canada, 1896–1930*, Canadian Studies in Economics, no. 2 (Toronto, 1955), pp. 19–25.

and the manifold processes of capital formation on the frontier, all these were familiar phenomena. The wheat economy was apparently again on the road to rapid expansion.

Similarities between the late 1920's and former periods, however, were more apparent than real. Or, more precisely, individual activities were very much the same as they had been before the war, but their collective significance was vastly different. Collectively the economic processes associated with the wheat frontier had been of a magnitude before 1914 sufficient to vitalize and integrate the entire Canadian economy and to diffuse this economic vitality throughout the North Atlantic trading area. In absolute quantities the immigration, occupation and improvement of land, and capital formation which were associated with the wheat economy in the latter half of the 1920's were smaller than before the war. In relative terms, however, the diminution was greater still, for other frontiers had meanwhile risen to prominence within the Dominion and the wheat frontier was no longer of unique importance. Possibilities for agricultural settlement in western Canada were no longer adequate to focus attention throughout the Dominion or to serve as the integrating force within the Canadian economy.

The new investment frontiers which became important in Canada in the first inter-war decade were associated with technological change and with new industrial demands. Improvements in hydro-electric technology, in the internal combustion engine, and in metallurgical processes gave significance to new resources and new areas. The manufacture of pulp and paper and the extraction of non-ferrous metals assumed increasing importance. Exports of these products amounted to 30 per cent of total Canadian exports in 1929 as compared with 19 per cent in 1920.[15] Hydro-electric power provided the essential energy. The developed water power in Canada increased from 2.5 million to 5.7 million horsepower from 1920 to 1929.[16] The raw materials for the paper and mineral industries and the power sites indispensable for their utilization lay in convenient juxtaposition in the Shield in northern Ontario and Quebec and in the mountain valleys of lower British Columbia. The manufacture of newsprint, the extraction of base and precious metals, and the installation of hydro-electric power facilities are extremely dependent on capital. The opening of frontiers based on these activities involved, therefore, investment and capital formation on a lavish scale. In the decade of the twenties at least $6 billion were invested in capital goods

[15]*Report of the Royal Commission on Dominion-Provincial Relations*, Book I, p. 116.
[16]*Ibid.*, n.

in Canada.[17] Of this sum, upwards of one-quarter went into central electric stations, pulp and paper mills, smelters, and metal refineries.[18]

The geographic diffusion of investment frontiers in Canada after 1920 is of significance for Dominion-provincial relations and for the national policy. Provincial control of resources as provided for by the British North America Act meant comparatively little for several decades after Confederation. It became a matter of some interest over the turn of the century with the early recognition of tentative prospects in minerals and in the production of pulp-wood for the export market. By the end of the First World War, with these tentative prospects tested and matured to the point where they might serve the leading economic requirements of the country, the certainty of provincial responsibility was a matter of prime importance. Control of economic activities on the Canadian investment frontier was no longer exclusively a federal matter. The Dominion government continued to administer land policy in the prairie provinces, presumably for the purposes of the Dominion. Elsewhere, however, each provincial government administered lands and resources for the purposes of the province. The constitutional diffusion of the economic frontier in Canada after 1920 was as pronounced as was its geographical diffusion.

The governments of the prairie provinces had long urged that control of the natural resources within their boundaries should be transferred to them in order that the anomaly of differing levels of provincial status might finally be removed. Furthermore, they demanded compensation for resources already alienated so that the grant of full provincial status would, in effect, be retroactive. The principles embodied in these claims were eventually recognized by the Dominion government in the later 1920's and royal commissions were appointed to work out the terms of settlement. The terms as recommended and adopted provided for adjustment retroactive in each case to the province's entry into Confederation. This meant, of course, 1870 for Manitoba and 1905 for Saskatchewan and Alberta.[19] Details of the transfer were near completion by 1930. The federal government was ready to concede that the remainder of the agricultural land and other natural resources in the prairie provinces no

[17]Ibid.

[18]Ibid. An incidental point of some significance is that the investment boom of the 1920's was largely financed out of Canadian funds. Of the $800 million to $900 million net long-term capital imports into Canada during the twenties, however, approximately two-thirds came from the United States. Ibid.

[19]The Saskatchewan government contended that the adjustment should be made retroactive to 1870 but, on a stated case, the Supreme Court of Canada and the Privy Council ruled that the province's claims could not predate September 1, 1905. See ibid., p. 135.

longer concerned the purposes of the Dominion but rather those of the individual provinces.

The transfer of natural resources to the prairie provinces in 1930 coincided with the commencement of a decade of unprecedented economic distress in the Canadian West. The economic disaster which encompassed the wheat economy at that time was compounded from the ill effects of world-wide depression and persistent local drought. These forces halted the economic expansion of the wheat economy and replaced material advance with retrenchment and widespread destitution. Investment gave way to massive and prolonged disinvestment. Relief was called for on a scale far beyond the competence of municipal and provincial governments. Only the Dominion government was adequate to the task.

The details of the disastrous thirties, or of the war and post-war circumstances of the forties, are not at present our concern. The question which was in doubt in the early part of the 1930's was whether the expansion of the wheat economy on a significant scale was at an end or only temporarily interrupted. The distant future cannot be surmised; but Canadian wheat acreage in the middle 1950's is approximately equal to that of the early 1930's. Depression and drought have spared the Canadian wheat economy for a period of more than fifteen years. Yet that economy has not appreciably extended its sphere. It appears increasingly appropriate to recognize the transfer of natural resources to the prairie provinces in 1930 as marking the end of chapter one of the national policy. Whatever our terminology, it is clear that the end of the 1920's marked the end of an era in Canadian economic and constitutional life.

PART TWO

GRAIN MARKETING PROBLEMS IN WESTERN CANADA 1900-20

→»)«←

CHAPTER SIX

The National Policy:
Competition and Monopoly

WHEAT is the second great staple product derived from the interior continental plain. The twentieth-century grain trade is the modern successor to the fur trade which formed the economic basis of the prairie economy from 1670 to 1870. The wheat grower succeeded the Indian as the "native" producer of the staple product sought after by urban populations whose demands create the opportunity for commercial profits. Governmental policy concerning the market structure of the wheat economy was not formulated in an historical vacuum but was founded on two traditions. There was, on the one hand, the long-established attitude of government toward staple production and the marketing structure in the interior as represented by the relationship between both imperial and colonial administrations and the Hudson's Bay Company: the Company held the bulk of its trade under charter-granted monopoly and governments were content to regard its monopolistic position as natural and perpetual. On the other hand, there was the complex of attitudes regarding agriculture and agricultural development which emerged with the national policy.

The fur trade was closely integrated and monopolistic. The native producer of fur faced a high degree of monopoly in both his selling and his buying markets. In the market in which he sold his furs he faced a single company, an absolute monopsonist. In the market in which he bought manufactured goods, "trade goods," he faced the same single company, an absolute monopolist, the sole source of supply for manufactured equipment. Both market transactions took place, of course, over the same counter and at the same time. In both markets the natives competed actively with each other. In the records of the fur trade there is no information known to the author which would suggest that the natives ever combined successfully to restrict purchases and thus to lower the price of manufactured goods, or to curtail sales and thus to raise the price of furs. In numbers they were apparently sufficient to fulfil the free competition requirement of "numerous sellers" in the selling market and of "numerous buyers" in the market in which they bought manufactured goods.

There were, of course, periods of substantial international or inter-regional competition within the fur-trade system. The Hudson's Bay Company, clinging to its original tidewater trading posts, faced the vigorous competition of French traders from the St. Lawrence and eventually established posts inland in order to counter this competition. For fifty years after 1770 they faced still more strenuous competition from the Anglo-Americans from Montreal. The latter had a good deal of trouble in stifling competition among themselves and found at least a partial solution for their difficulties by expanding into the territory of the Hudson's Bay Company. Amalgamation of the Hudson's Bay and North West companies in 1821 established the former as absolute monopolist in the merchandising field—both buying and selling—in western Canada.

Although the monopoly of the merchandising interests in the North American fur trade came under serious threat from time to time, it was the threat of business rivals rather than of governmental interference. The Hudson's Bay Company held its control over the economic activities of Rupert's Land under a royal charter which continued in effect for two centuries without effective legal challenge. The pattern of economic imperialism by chartered monopolistic trading companies, well set in the western European tradition after the creation of the great East Indian companies at the beginning of the seventeenth century, was applied to the mid-continent of North America after 1670 and was maintained beyond the middle of the nineteenth century.

The first successful challenge of the privilege of monopoly in the Canadian West came by way of local agrarian revolt rather than by governmental curtailment. The incident concerned the Hudson's Bay Company and the Red River colonists and, in its various aspects, strikingly foreshadowed later events related to monopolistic positions.

Between 1821 and 1850 the Red River settlers clashed frequently with the Hudson's Bay Company on the matter of trading rights. On granting Assiniboia to Lord Selkirk, the Company had reserved for itself all trading rights and had specified that neither Selkirk nor his colonists should "carry on or establish . . . any trade or traffick, in or relating to any kind of furs or peltry."[1] In the very early years of the colony's existence the residents traded with the Indians for furs and, in turn, with American fur traders in Minnesota. From the early 1820's until 1832, when Assiniboia was returned by the sixth Earl of Selkirk to the Hudson's Bay Company, freedom of trade was for the most part

[1]As cited in J. P. Pritchett, *The Red River Valley, 1811–1849* (New Haven, Conn., 1942), p. 250.

unchallenged in the settlement. After the restoration of the territory in 1832, the Company set about enforcing its charter rights in regard to trade. It established tariffs and introduced espionage, open and armed search, confiscation of furs, and arrest and trial. It interfered with importation from England. For a time in the years 1846 to 1848 military government was in effect.[2] The settlers protested the Company's claim to a trade monopoly, before the local governor and to the British and the Canadian governments, but were accorded no satisfaction—the charter was legally sound. In the spring of 1849 a Métis, Sayer, and three others were brought to trial for illegal traffic in fur. The Métis organized in a determined effort to prevent, either peacefully or by violence, the conviction of the accused. On the day of the trial there were five or six hundred armed persons outside the court. Sayer was found guilty but was allowed to go without penalty. Although no commitment was made concerning future policy, the Métis celebrated the verdict as a sign that "le commerce est libre."

The Métis interpreted correctly the action of the court. The Hudson's Bay Company's trading monopoly in the settlement, although legally intact, was in practice at an end. During the 1850's the investigations conducted by special committees of the British and Canadian Houses of Commons indicated clearly that the legal residue of the monopoly, both territorial and commercial, would have to be liquidated if any economic or political development of the western territories were to be contemplated. This liquidation was not accomplished until after Confederation and then only by negotiation and by purchase of the Company's equity in the West.

The recognition that the Company's monopoly had outlived its usefulness coincided with the emergence of the national policy and might be taken to suggest the appearance of a critical attitude in government circles concerning the efficacy of monopoly. There is, however, no other trace of evidence to support this suggestion. Neither the factual liquidation of the Company's trading monopoly in the Red River settlement around the middle of the century, nor the negotiated cancellation of its exclusive proprietary and trading rights in 1869 marked a conviction in government circles that monopoly was necessarily undesirable or that it ought to be curbed by government action. The policy of the government was strictly empirical. The monopoly of the Hudson's Bay Company was a threat to the national purposes: therefore it had to go. Other monopolies, which did not threaten national purposes, were of no concern.

The Dominion government in fact perpetuated certain aspects of the

2*Ibid.*, p. 261.

chartered-company instrument of imperialism in its method of directing western development. Commercial monopoly or political sovereignty were no longer in question. But the Canadian Pacific Railway syndicate was granted a charter under Dominion legislation which, along with a considerable range of other privileges, extended protection against the construction of lines by other railways in certain areas.[3] Clause 15 of the contract between the government and the Company forbade for twenty years the construction of any line to the south or southwest of the Canadian Pacific Railway to reach within fifteen miles of the international boundary.

The "monopoly clause" of the contract came under attack from western interests even before it was enacted. When the charter was under discussion in the federal House in December, 1880, representatives of the province of Manitoba protested various of its clauses including clause 15. In reply to their criticism, Sir John A. Macdonald and Hon. Thomas White stated that the clause would not bar Manitoba from chartering a railway from Winnipeg to the boundary; it would only serve to prevent American railways from tapping Canadian Pacific territory *beyond* the boundaries of Manitoba.[4]

The discussion concerning the Canadian Pacific Railway Company's charter prior to its enactment was but the preliminary encounter in a series of bitter disputes which ended in the cancellation of the monopoly clause in 1888. Following the granting of the charter to the Canadian Pacific Railway syndicate, the Manitoba government chartered the construction of three railways out of Winnipeg, one to run southeast to the boundary, one to run southwest, and one with more general rights. The Canadian Pacific Railway Company protested these grants as an infringement of its rights. The Dominion government disallowed all three of the Manitoba charters, arguing that it was justified in barring the construction of any railway, even one entirely within provincial boundaries, if it connected with a foreign railroad, "a connection with foreign lines being contrary to the spirit of the British North America Act."[5]

The Manitoba people insisted that the policy of disallowing provincial railway charters be abandoned by the Dominion government. In 1884 Hon. Sir Charles Tupper, federal Minister of Railways and Canals, gave assurance that the government would comply with their wishes after the

---

[3]*Statutes of Canada*, 44 Vic., c. 1 (1881).

[4]Canada, *House of Commons Debates*, 1880–1, pp. 575–6.

[5]See Harold A. Innis, *A History of the Canadian Pacific Railway* (Toronto, 1923), p. 179.

completion of the Canadian Pacific Railway, but the situation remained uncertain. In 1887 two more bills incorporating railways in Manitoba were disallowed by the Governor in Council. Another one, making six in all, was disallowed in July, 1887, after work on the line had been started. In June, the Manitoba Legislature adopted a resolution of protest against what it claimed to be an arbitrary exercise of the veto power. In September, 1887, the Canadian Pacific Railway Company stated its position in regard to the rights which it claimed. The argument of the Company was that it needed protection in view of the long, unproductive milage which was necessarily a part of its system, and that the interests of the residents in Manitoba should not be allowed to interfere with those of residents in the older provinces. The Company threatened to move its main shops from Winnipeg to Fort William. With the Dominion government supporting the railway, the chances for a suitable compromise appeared slight. Professor Innis says: "Manitoba was insistent. The company refused to yield. Finally the Canadian Government gave way."[6] In April, 1888, the monopoly clause was cancelled by agreement between the federal government and the Canadian Pacific Railway Company, the government in return guaranteeing the interest on a new issue of bonds for the Company in the amount of $15 million.

The early experience of the Dominion government with the Canadian Pacific Railway Company confirmed that of the British government with the Hudson's Bay Company in indicating that a chartered commercial company with monopolistic powers was likely to be an ineffective instrument for the promotion of agricultural settlement. The inadequacy of commercial monopoly to the tasks of colonization had, indeed, been demonstrated thoroughly in the first half-century of French experience in New France. But the basic incompatibility between the fur trade and agriculture was readily understandable. Agricultural settlement was not only a costly burden for a fur-trading company to undertake in the first instance, but in proportion as such settlement prospered and grew, it threatened the production and monopoly marketing of furs on which the company depended for profits. The relation between a transportation company and agriculture would surely be different. What profit could a railway company derive from an agricultural region unless agricultural settlement expanded therein? It would have seemed impossible in 1880 that the newly chartered Pacific railway could have any interest which would not be compatible with the utmost expansion and prosperity of the western agricultural region. Particularly must this have been so when

[6]*Ibid.*, p. 182.

the agreement had provided for a grant of 25 million acres of land to be selected from areas in the West lying along the railway's right of way or elsewhere which should prove to be "fairly fit for settlement."

The Canadian Pacific Railway Company has always argued that its own best interests must obviously lie in the pursuit of policies which would foster western expansion and prosperity. After 1900 the Company was the leading partner of the Dominion government in the active and successful promotion of immigration and agricultural settlement in western Canada. Yet this was only after the Dominion government had been driven to cancel the Company's monopoly. The experience of the Dominion government and the settlers of Manitoba with the Canadian Pacific Railway Company, while the monopoly clause remained in force, indicated that the Company was prepared to discourage western expansion rather than have it take place at a rate or in directions that would threaten the Company's control and ability to secure whatever profit there might be.

The cancellation of the monopoly clause of the Company's charter—cancellation, in effect, on a purchase basis—was basically a matter of opposing pressures, but in retrospect it can be seen to have had various implications. It indicates the recognition by the government, first, that the Canadian Pacific Railway Company was an instrument of national policy rather than an end in itself and, second, that the benefits of the expansion contemplated in the national policy were not to be regarded as the narrow game preserve of a select few of the eastern interests but were rather to be multiplied and widely diffused. There were many eastern interests. For example, in addition to the Canadian Pacific Railway Company, which was the leading transportation interest, there were eastern manufacturers of all sorts; there were importers and wholesale merchants of national importance; and there were other banks besides the Bank of Montreal. There were other commercial and financial centres besides Montreal, none with her economic strength, no doubt, but some nevertheless possessed of a cumulative economic and political significance which was far from negligible. These would include Toronto, Quebec City, Halifax, and perhaps one or two others. Consideration of these other interests was not a fundamental factor in the decision to cancel the monopoly clause in 1888. It was, however, of basic importance in determining the attitude of the Dominion government on railway matters after 1900, when the government came not only to tolerate but actively to promote the construction of two additional transcontinental railways.

Another implication of the cancellation was the suggestion that there might henceforth be circumstances in which the views of westerners on policy for the West would be found more acceptable to the federal administration than the views of any particular group of eastern interests. This eventuality was difficult for easterners to contemplate. Europeans on the St. Lawrence had been accustomed in the fur trade to dealing with native producers whose language and cultural background differed vastly from their own. The natives' views on matters of importance were of complete indifference to the traders. It is not too much to suggest that something of this arrogance was carried over into the methods of fostering the development of the western agricultural economy. But the staple-producing "natives" of the wheat economy were themselves of European stock; the majority of them were Anglo-Saxon and many of them came directly from eastern Canada. They voiced their views in forthright English with every confidence that these views were worthy of attention.

The cancellation of the monopoly clause may have been among the first significant modifications in federal policy to result from the advice and stubborn pressure of westerners. It was by no means the last. After 1900 there were many such modifications. In many instances, however, the West applied pressure to no avail; policy had all the appearances of immutability. The distinction between elements of policy which were changeable under western pressure and those which were not goes back to the distinction between the two basic economic elements of the national policy in so far as it concerned the West. These elements were, first, to encourage the maximum economic development of the western territories and, second, to assure the integration of this development into the national economy. Western advice which appeared to concern western development—and if it did it was most unlikely to favour the retardation of that development—was often quite acceptable to eastern policy-makers. Western advice which related to the question of national integration, however, was seldom found acceptable because it was most likely to be opposed to integration on eastern terms. The prominence and acceptability of western views in the formulation of grain trade regulations after 1900 is an example of the first situation. The persistent rejection of western opinion in the matter of the tariff is an illustration of the second.

The Canadian Pacific Railway Company's monopoly was, of course, a special case, and it had been granted by the Dominion government under special agreement with the Company. The cancellation of this privilege gives no indication of the policy of the Dominion government on the question of monopoly in general. The western settler encountered many other groups in the market place in addition to the Canadian Pacific Railway Company, and his economic well-being was affected by the degree of competition or monopoly which characterized each of these groups. If he found himself faced with monopolistic buyers when he attempted to sell his produce, or with monopolistic sellers when he went to buy the equipment and consumers' goods which he required for his farming operations, what was he to expect from the Dominion government by way of action to assure competitive equality?

The interest of western Canadians in the problems created by monopoly in the late nineteenth century was strengthened by the movement of protest against monopoly in the United States. The Granger protests in the mid-western states in the 1870's and the state legislation which followed were particularly concerned with railway monopoly and railway rates and services. The Interstate Commerce Commission was established by the federal government in 1887 to deal with these problems on an interstate basis. The Sherman Antitrust Act of 1890 was couched in terms which were presumably broad enough to abolish monopoly wherever and whenever it might be found. Legal action against American trusts was intensified over the turn of the century and was signalized by successful prosecution in the Northern Securities case in 1904. The prevalence of industrial and commercial combination and the effect on society of irresponsible business, large and small, were constantly publicized by the "muck-raking" journalists of the period.[7]

The threat of monopoly aroused considerable concern in Canada in the decades after Confederation. In 1888, Clark Wallace, Conservative member from Toronto, secured the appointment of a committee of the federal House to investigate charges that combinations existed in various Canadian businesses. After two months of sittings, the committee reported that combines existed in the manufacture of sugar, coffins, bindertwine, stoves, oatmeal, biscuits, and confectionery, as well as among certain wholesalers and retailers.[8] Wallace introduced a private member's bill for the suppression of combinations in restraint of trade. Tabled for a year, this bill emerged in modified form in 1889 but was reduced to practical impotence by the Commons and the Senate before its enactment. During the 1890's only one case arose under the act, an action against the American Tobacco Company, and the Company was cleared. This legislation was strengthened somewhat in 1900 and was

[7]For an interesting analysis of this aspect of American journalism over the turn of the century see C. C. Regier, *The Era of the Muckrakers* (Chapel Hill, N.C., 1932).

[8]Lloyd G. Reynolds, *The Control of Competition in Canada* (Cambridge, Mass., 1940), p. 133.

incorporated in the Criminal Code as sections 496–498.[9] Six cases came under these sections from 1900 to 1923. In 1897, amendments to the Customs Act provided that if investigation indicated the existence of a combination in the production of a protected article, the Governor in Council might place the article on the free list. This act was applied in only one case, that of newsprint.

The control of combination in the Canadian business world has been dealt with almost exclusively under the terms of the Combines Investigation acts of 1910 and 1923 with amendments to the present time. Both acts were introduced and sponsored by the late Rt. Hon. W. L. Mackenzie King. The philosophy underlying this legislation is that control of business practices can best be secured by investigation and publicity. The act of 1910 was totally ineffective and was called into operation but once. The act of 1923 was a considerable improvement. Under its terms, however, which remained substantially unchanged until 1952 (most of the amendments have concerned administrative detail primarily), illegality existed only where business groups of various specific sorts "have operated or are likely to operate to the detriment or against the interest of the public, whether consumers, producers or others."[10] An amendment in 1951 prohibited resale price maintenance. Legislation in 1952[11] provided the first thorough revision of Canadian combines legislation since 1923.

Between 1923 and 1940 twenty full investigations were held under the Combines Investigation Act and combines were found in nine cases. Six prosecutions resulted in five convictions. Of the twenty investigations, one, that into the purchasing of tobacco, was of direct concern to Canadian farmers, though none was of immediate significance to western farmers. During the 1940's there was some increase in the activities of the Combines Investigation Branch of the Dominion government, and at least two of the investigations were of more immediate concern to the

[9]Ibid., p. 136.

[10]Revised Statutes of Canada, 1927, c. 26.

[11]Statutes of Canada, 1 Elizabeth II, c. 39 (1952). On the administrative side, the amendments of 1952 revised procedures so as to separate the functions carried on under the act into (a) investigation and research, and (b) the appraisal of, and reporting upon, evidence derived from investigations. The commissioner provided for under the former legislation becomes the director of investigation and research. He has authority to initiate investigations of practices falling under the Combines Investigation Act or under sections 498–498A of the Criminal Code when related to practices covered by the Combines Investigation Act. The appraisal of evidence and the reporting thereon now fall to a board of three members, the Restrictive Trade Practices Commission. Other changes of 1952 removed former limitations on maximum fines and permitted the courts to prohibit the continuation or repetition of an offence.

western Canadian farmer.[12] These, however, lie beyond the scope of the present study. The Combines Investigation Act of 1923 was one of "eight important statutes" passed in Canada between 1899 and 1940 concerning the regulation of monopoly, but Professor Reynolds points out that "nearly all the effective action in this field" before the Second World War was taken under the act of 1923.[13]

A fair inference from the available information is that action concerning combination and monopoly has been of little concern to the Dominion government. There has been no "trust-busting" era in Canada as there was in the United States—although American experience has by no means proved the efficacy of such activity. The respective titles of the principal acts in the two countries are symbolic. The Canadian legislation has been called Combines *investigation* legislation; the American counterpart was originally the Sherman *Antitrust* Act. Of even more importance, however, in indicating and explaining the comparative leniency of Canadian legislation is the fact that it has never been enforced with marked rigour. Neither of the major Canadian political parties has developed, or found it expedient to develop, a conviction that monopoly in itself is bad, or that, if the actions of a particular monopoly were questionable, political advantage would assuredly follow the selection of that monopoly for treatment by way of example.

A proper appreciation of the place of the Combines Investigation acts in Canadian jurisprudence requires not only the recollection that they were formulated and sponsored by the late Rt. Hon. W. L. Mackenzie King, but also that the thirty-year period after the end of the First World War can properly be regarded as the "King era" in view of Mr. King's domination of Canadian political thought and action throughout that span of years. His social philosophy is apparent, then, not only in the terms of Canadian combines legislation but also in its application to the Canadian business world. Writing on this point in 1940 Professor Reynolds said:

Mr. King . . . is a liberal in the British tradition. He prefers to leave business men to run their affairs, though it may occasionally be necessary for government to intervene as a mediator or to protect the interests of defenseless groups. The growth of large-scale enterprises should be encouraged, he

---

[12]E.g., *Bread-Baking Industry in Western Canada: Report of H. Carl Goldenberg, Special Commissioner, of an Investigation into an Alleged Combine in the Bread-Baking Industry in Saskatchewan, Alberta and British Columbia* (Ottawa, Nov. 3, 1948); and *Flour-Milling Industry: Investigation into an Alleged Combine in the Manufacture, Distribution and Sale of Flour and Other Grain-Mill Products* (Ottawa, Dec. 29, 1948).

[13]*The Control of Competition in Canada*, p. 169.

believes, and amalgamation of business units is in most cases beneficial. Monopolies and trade agreements must be carefully watched, however, to see that they do not abuse their economic power. Mr. King believes strongly in the preventive value of publicity and considers that the possibility of investigation will do more to deter potential wrong-doers than any amount of criminal investigation. His speeches in Parliament on the two Combines acts are able expositions of these beliefs.[14]

The above review of governmental actions and attitudes suggests that the Dominion government has not concerned itself to any extent with removing the inequalities of bargaining power associated with the existence of monopolies. This statement is valid at least for the period throughout which the national policy was introduced and established. The Dominion government formulated and developed the national policy in accordance with an economic philosophy identical with that which prevailed in the provinces for whose benefit the new policy was created. The federal government was originally, and long remained, convinced of the worth of free enterprise. The one major compromise was the admission that private enterprise could not be expected to undertake the burdens entailed by transportation improvements in a new continent and by the revolutionary new techniques of the railway age. Freedom of enterprise, held to be desirable by successive Dominion governments regardless of their political views, has meant freedom to compete and, apart from exceptional cases, freedom to combine; it has meant freedom to grow large and economically powerful or to remain small and weak.

Since competitive inequalities have ordinarily been fully compatible with the economic philosophy of the Dominion government it becomes important to indicate the extent to which such inequalities have characterized the market activities of the prairie farmer.[15] In addition to a great variety of information which has been placed on record from time to time since the turn of the century, sometimes in investigations of agricultural conditions and sometimes in other investigations, a great deal

[14]*Ibid.*, p. 137 n.

[15]The interpretation of facts rather than the facts themselves constitutes the basis of agrarian attitudes toward the market place. The following statement typifies the farmer's view of his market experience: "The summary of the farmer's grievance is that he must exchange his products at prices determined by the rigid economic law of supply and demand, or lower, for services and commodities at prices fixed by agreements, combines and monopolies, to which, although undoubtedly exorbitant, the law gives its benediction and approval." From a letter to the editor, by P. D. McGrath, Cherhill, Alberta, in the *U.F.A.*, June 15, 1927, p. 32. The influence of such views in the formulation of the objectives of farmers' organizations can be clearly traced in the following statement made by a vigorous advocate of the Canadian wheat pools: "What would you think," he asked, "of

of information was recorded in the 1930's in the course of the economic soul-searching of that desperate period.[16]

Basing his analysis chiefly on evidence adduced before the Price Spreads Commission in the years 1934 and 1935, Professor Reynolds presents an excellent summary view of large segments of Canadian industry in the middle thirties. His analysis is not specifically directed toward agriculture but it throws a good deal of light on the agricultural situation, because many of the manufacturing industries which he deals with either produce commodities of importance in the farmer's buying budget or process farm products and thus constitute important intermediate markets. It will be necessary to examine some of the non-agricultural areas of the Canadian economy in greater detail and with regard to their changing circumstances in the decades after 1900. But allowing for historical changes, which cannot be traced in detail without exhaustive research, the summary view of conditions in the 1930's will serve to typify many of the market relationships encountered by prairie farmers in the decades during which the wheat economy was taking shape.

Professor Reynolds finds that the most significant characteristic of Canadian industry in the middle thirties was its "concentration of output." "The most striking fact about Canadian manufacturing," he says, "is the fewness of producers in most fields. In most industries there are not more than ten important producers, and in many there are only three or four."[17] Among the most important of the factors which accounted for this situation Professor Reynolds lists the following: (1) the smallness of the Canadian population and of the Canadian market; (2) the splitting of the domestic market into four distinct re-

---

a manufacturer of some indispensable product, were he to go among his potential customers and allow them to set his market price, for something which they need, and which has cost him a specific sum to produce? It simply isn't done. Yet there stands the farmer, with his eternal morale-smashing inquiry: 'What are cattle worth?'—'Do you want any eggs?'—'Are you in the market for hogs?'—'What are you paying for potatoes?'—'What happened to wheat today?' " Walter P. Davisson, *Pooling Wheat in Canada* (Ottawa, 1927), pp. 85–6.

[16]See particularly *Report of the Royal Commission on Price Spreads* (Ottawa: King's Printer, 1937); House of Commons, session 1937, Special Committee on Farm Implement Prices, *Minutes of Proceedings and Report* (Ottawa: King's Printer, 1937); Legislature of the Province of Saskatchewan, session 1939, Select Special Committee on Farm Implement Prices and Distribution, *Report* (Regina: King's Printer, 1939). See also Combines Investigation Act, *Investigation into an Alleged Combine in the Bread-Baking Industry in Canada: Report of Registrar, February 5, 1931* (Ottawa: King's Printer, 1931); and *Report of the Royal Commission on the Textile Industry* (Ottawa: King's Printer, 1938).

[17]*The Control of Competition in Canada*, p. 40.

gional markets by distances and transportation costs; (3) the reduction of the number of firms through consolidation; and (4) the extension of many of the larger American producers into Canada. Concluding his remarks on the latter point he says: "Canadian Celanese, Canadian Westinghouse, Canadian General Electric, Aluminum Limited, International Harvester, the 'big three' automobile companies and the 'big four' tire producers occupy the same dominant position in Canada that their parent companies enjoy in the United States."[18] He also notes that, while a number of the American companies were more or less successfully modified in the United States under the Sherman Act, "their Canadian offspring led an untroubled life in Canada and still dominate their respective industries."[19]

Professor Reynolds compiled a table summarizing the available information pertaining to concentration in Canadian industry in the middle thirties. This table is adapted and included here as Table V.

TABLE V

CONTROL OF OUTPUT OF SELECTED INDUSTRIES IN CANADA IN THE MIDDLE 1930's

Industry	Cumulative percentages of output controlled by				
	1 firm	2 firms	3 firms	4 firms	5 firms
Automobiles	40	65	89	—	—
Agricultural implements	—	—	—	75	—
Cement	90	—	—	—	—
Copper	53	—	—	93	—
Canning (fruit and vegetable)	67	83	—	—	—
Cotton yarn and cloth	48	—	79	—	—
Electrical equipment (heavy)	—	—	100	—	—
Fertilizers	—	—	70	—	—
Lead	91	—	—	—	—
Meat packing	59	85	—	—	—
Milling	—	—	—	—	73
Nickel	71	—	—	—	—
Oil	55	—	—	—	—
Rubber footwear	39	50	61	72	—
Silks, real	23	42	61	—	—
Silks, artificial	66	100	—	—	—
Sugar	—	—	—	—	100
Tires	—	—	—	—	65
Tobacco	70	90	—	—	—
Zinc	74	—	—	—	—

SOURCE: Adapted from Table I in L. G. Reynolds, *The Control of Competition in Canada* (Cambridge, Mass., 1940), p. 5, and printed here with the kind permission of the author and publisher. Professor Reynolds points out that the figures refer to volume of output or of sales in a given year, most of them in the year 1933 but a few of them in 1934 or 1935.

18*Ibid.*, p. 7. Professor Reynolds mentions Imperial Oil, incorporated in 1880 as a subsidiary of Standard Oil, and Imperial Tobacco Company, incorporated in 1895 as the American Tobacco Company of Canada, as other outstanding examples of the extension of large American producers into Canada.

19*Ibid.*

The degree of concentration illustrated in Table V is to a considerable extent the result of a consolidation movement which has long been under way in the industrial field. The Price Spreads Commission pointed out that there were 374 consolidations in Canadian industry in the years 1900 to 1933 and that these consolidations absorbed 1,145 concerns.[20] The progress of the trend toward consolidation was not regular, but was concentrated mainly in two periods: the first from 1909 to 1912, inclusive, during which 58 consolidations occurred, and the second from 1925 to 1930, during which 231 occurred. In the earlier period, important moves were made toward concentration of control in the heavy industries such as cement, asbestos, coal, and iron and steel, as well as in textiles, milling, pulp and paper, and others. The consolidation movement spread particularly widely in the latter half of the 1920's and drastically altered the competitive position in such industries as pulp and paper, milling, baking, brewing and distilling, meat packing, canning, and the processing of dairy products. On the question of purpose Professor Reynolds says, "The evidence is . . . that the dominant motives in these consolidations were the desire for price control and for promoters' profits."[21]

Concentration in industry is significant because of the fact that, as a general rule, the possibility of restricting competition varies inversely with the number of producers involved. The discrepancy between industry and agriculture is particularly striking when the conditions outlined in the preceding paragraphs are contrasted with the numerical facts applicable to farm production. There were 200,000 farms in the prairie provinces by 1911 and an all-time maximum of 300,000 by 1936. It is obvious that there can be no agreement on price or output among producers who compete in the market in such vast numbers. Not every western farm produces wheat exclusively or even primarily, but wheat is the chief product of by far the greatest proportion of the total and has been ever since western agricultural development began.

---

[20]*Report of the Royal Commission on Price Spreads*, p. 28. Consolidation of individual firms does not, of course, constitute the only restraint upon competition. Price agreements and trade association may be very effective and were well established in Canada before the turn of the century. In 1905 a raid on the offices of Jenkins and Hardy (Toronto accountants and professional trade association secretaries) ". . . gave evidence of price agreements covering (among other things) axles, churns, cordage, harness, handles, wheels, edged implements, springs, lanterns, stamped metal ware, saws, trunks, tacks, wringers, washing machines, wooden-ware, oatmeal, canned goods, lead pipe, soil pipe, plate glass, screws, bolts and wire nails." Reynolds, *The Control of Competition in Canada*, p. 17 n. Professor Reynolds points out that no trials followed these disclosures.

[21]*The Control of Competition in Canada*, p. 6.

Considering agriculture as a whole, whether we have in mind the wheat farmer, the livestock or dairy producer, the fruit or vegetable grower, or the producer of specialty crops such as wool, tobacco, honey, or fur, it is a rare instance where there are not hundreds or thousands, even hundreds of thousands, of farmers competing in the same market with the same product. In such circumstances, price agreements, restrictions on quantity, orderly marketing—any of the paraphernalia effective in the suppression of competition, in fact—are beyond the reach of producers. In contrast to the market position of farmers is that of the producers of many of the goods and services which farmers buy, where production is typically concentrated in a few hands and where the control of output and price is therefore relatively easy.

The fact that there are so many farmers in each line of agricultural production and so few producers of many of the goods and services which farmers buy is no accident. The reason lies in the economies of large-scale production, in the fact that unit costs in every line of economic activity tend to be reduced if the scale of production is increased up to a point, and that they tend to rise if the scale is further increased. In some lines of production the optimum-sized plant, the plant capable of minimizing unit costs, is an extremely large concern; in others it is very small. Most manufacturing and processing industries, as well as transportation and warehousing concerns, are in the former category; agriculture falls in the latter.

Closely associated with numerical disparities and the corresponding inequality in the degree of effective competition as between industries is the relative responsiveness of various industries to the pressure of changing economic circumstances. The onset of depression finds a highly competitive industry in an extremely vulnerable position; the producers

TABLE VI

PRICE AND PRODUCTION INDICES FOR AGRICULTURE AND
AGRICULTURAL IMPLEMENTS IN CANADA, 1929–33

	AGRICULTURE		AGRICULTURAL IMPLEMENTS	
	Prices	Production	Prices	Production
1929	100.0	100.0	100.0	100.0
1930	81.6	101.9	98.8	72.8
1931	55.9	97.9	98.8	19.2
1932	48.0	113.1	97.3	9.0
1933	50.6	104.0	93.6	13.6

SOURCE: Dominion Bureau of Statistics, as cited in *Report of the Royal Commission on Price Spreads* (Ottawa: King's Printer, 1937), p. 9.

have little ability to curtail output, so the market pressure works itself out in a price decline. In an industry which is monopolistic or at least monopolistically competitive, price declines may be averted or minimized by restriction of output. Table VI illustrates these tendencies in the depression years of the early thirties.

Prices of agricultural products in Canada fell 50 per cent from 1929 to 1933 while production *increased* 4 per cent. In sharp contrast, prices in the agricultural implement industry declined less than 7 per cent while physical production was *reduced* by as much as 91 per cent.

CHAPTER SEVEN

Local Elevators and the Monopoly Issue before 1910

THE FARMERS IN THE PRAIRIE PROVINCES protested repeatedly, if not continuously, against their position in the price system throughout the decades which marked the establishment and development of the Canadian wheat economy. They charged monopoly and extortionate pricing. They attacked "the interests" and the financial domination of eastern capitalists. They proposed far-reaching reforms. For certain industries they urged state ownership or state control. For others they demanded the removal of the tariff wall behind which they saw concentration and monopoly emerge. In particular cases they sought to combat monopoly by co-operative organization and production. Nowhere did the varied protests and proposals for reform come into sharper focus than in reference to the local market centre and the commercial agencies therein located. High and stable prices for the things they had to buy; unstable and often low prices for the products they had to sell—these, the farmers argued, were the terms of a formula for unpredictable and inadequate agricultural income.

The farmer's *monetary income* depends upon the quantity of produce which he has for sale and the price at which he sells it. His *real income* depends not only on the quantity and price of the produce which he sells but on the prices of the goods which he buys as well. Aware of this dual dependence, the prairie farmer has directed his attention partly to the possibility of increasing the selling prices of his products and partly to the possibility of lowering the prices of his purchases. Until recent years at least, however, efforts to reduce purchase prices have been sporadic and comparatively ineffective. Even at the present time, agricultural attack on the prices of farm supplies is effective only within a narrow range although concerned with important supplies such as tractor and farm fuels, feeds and fertilizers, building materials, and agricultural implements.[1]

[1]For a brief sketch of the status of co-operative farm supply in Canada see V. C. Fowke, "Developments in Canadian Co-operation," *Journal of Farm Economics*, Proceedings Number, Nov., 1951, pp. 909–17.

While the effects on real income of buying and selling prices are both readily discernible, it is nevertheless true that producers are more likely to show concern over the latter than the former. This is understandable because the producer ordinarily sells far fewer kinds of goods or services than he buys. A labourer, for example, sells but one product, his services, and has but one selling price to concern him. With the money secured from the sale of these services at a single, hourly, daily, or piece-work price he buys hundreds of individual goods and services at as many individual prices. Any hope of controlling or of adequately modifying the latter diffused price phenomena by individual effort must appear slight indeed. An individual farmer may produce and sell a variety of products, but never in numbers beyond the slightest fraction of the number he must buy. For the highly specialized agricultural producer such as the prairie wheat grower, cash income may almost literally be a function of the price of a single annual product, wheat. There is accordingly no occasion for surprise in the persistence of the efforts of prairie farmers, decade after decade, to establish effective control over the price of wheat.

Although there are several regional prices for Canadian wheat, including the Liverpool or "world" price and the Winnipeg or "spot" price, the one that concerns the farmer is the one obtainable at the local market centre. Under open market conditions, the price at any local point in the prairie provinces depended on the price at Liverpool and the costs of moving grain from the local point to Liverpool. The local price might therefore be raised by an increase in the Liverpool price, by a reduction in the costs, or by both together. Only by means of some plan of major proportions could there be any hope of modifying the Liverpool price. The spread, however, was a different matter, and, prior to 1920, prairie farmers were widely convinced that it was excessive. It was not a fixed amount but was made up of the freight, insurance, and handling charges, and might include a highly variable margin of profit for the grain handlers as well. The possibility of reducing the spread was rendered promising by the fact that certain of the agencies that gave rise to the costs operated within the wheat growing area and within full view of the agricultural producer. By far the most important of these were the elevator companies that assembled and stored wheat in the local centres and, after purchase for cash or under instructions for forwarding, loaded it in railway cars for movement to terminal markets.

The seasonal nature of the production and maturity of Canadian wheat, the seasonal features of the inland waterway comprised in the

Great Lakes, and the mid-continental location of the wheat growing areas have all combined to impose a colossal burden of capital requirements on the wheat economy for adequately performing the physical transfer of the Canadian crop. The farmer's need for maximum cash receipts, as soon as possible after harvest and threshing, creates demands for transfer and storage facilities at local points, for rolling stock and locomotive power to rush grain to the lakehead, for storage and transfer facilities in lakehead terminals, and for tonnage to move the grain quickly down the lakes. Adequate facilities are obviously required for the further stages of the movement, but it is not necessary to consider the process at greater length. It is only necessary to call attention to the interdependence of the various stages of the process and to point out that congestion at any point or in any stage of the marketing movement quickly spreads throughout other sections of the channel. In many of the crop years during the three-quarters of a century since western wheat first moved into the export market,[2] for example, a shortage of box cars and engine power served to "plug" the local elevators in many of the wheat producing areas of the West.

Canadian wheat, with the exception of that retained on the farm for seed, is normally disposed of to yield cash income. The physical transfer takes place at the local market centre and typically through a local ele-

---

[2]The first shipment of wheat from the Canadian West took place in 1876, before the coming of the railway, when R. C. Steele of Toronto, a founder of the Steele, Briggs Seed Company, came to Winnipeg to purchase seed wheat. He wished to secure 5,000 bushels but was able to purchase only 857. The wheat was sacked and shipped south from Winnipeg by steamer on the Red River to Fisher's Landing in Minnesota. Steele secured another 4,000 bushels of wheat in northern Minnesota and the total shipment was taken by rail from Fisher's Landing to Duluth, by boat to Sarnia, and by rail to Toronto. The growers in the Red River settlement were paid 85 cents per bushel and the freight to Toronto was 35 cents per bushel. Shipments of wheat to Great Britain, the first *exports* of Manitoba wheat, were made in 1878 and 1879. These shipments were carried up the Red River to go forward by American railways. Winnipeg and Minneapolis were linked by rail in 1879. With the completion of the Canadian Pacific Railway from Port Arthur to Winnipeg in 1883, James Richardson & Sons shipped the first cargo of wheat—one of 10,000 bushels—from the Canadian Lakehead to Owen Sound. The Canadian Pacific Railway Company began the construction of a terminal elevator at Port Arthur in 1883 and completed it early in 1884. In 1884 a sacked shipment of 1,000 bushels of No. 1 Manitoba Hard wheat was carried over the all-Canadian route to seaboard and thence to Glasgow, apparently inaugurating the direct export of Manitoba wheat by means of an all-Canadian route to seaboard. The decade 1876–85 saw the establishment of western Canada as a regular exporter of wheat. See D. A. MacGibbon, *The Canadian Grain Trade* (Toronto, 1932), pp. 25–8; and E. Cora Hind, "A Story of Wheat," reprinted from *Canadian Geographical Journal*, Feb., 1931.

vator. The cost[3] of providing adequate facilities for the storage and transfer of grain, and the shortage of capital, led, in the early days of settlement, to reliance on equipment vastly inferior to standard elevators, but which could be constructed with the resources available.

Flat warehouses were one alternative. They were essentially sheds which housed low bins. They were without elevating equipment or power, and grain could be loaded in or out, or moved from bin to bin, only by hand-powered shovels and trucks. They permitted farmers to accumulate grain at the siding in carload quantities in anticipation of shipment by rail, but apart from their cheapness of construction and the fact that they provided shelter from the elements they had little to commend them. They were a temporary makeshift, and disappeared from local market centres in the West at the first opportunity.

The loading platform was another alternative. It was simply a timbered structure with earthen ramps, erected by the railway companies beside the local siding on a level with the floor of a box car which might be "spotted" alongside, so that the grower could load grain into the car directly from his wagon without putting it through an elevator. The loading platform constitutes standard equipment at railway sidings to the present time. Its provision at every local point where a farmers' petition requested it was made compulsory in Dominion legislation for the regulation of the grain trade at the turn of the century.[4] Over the years the loading platform provided the shipper of grain in carload quantities a choice in the disposal of his crop. To ship thus to the terminal market freed him from reliance on the local elevator, it saved the handling charges, and it assured the preservation of the identity of his grain.[5]

[3]MacGibbon stated that the cost of constructing a country elevator with a capacity of 30,000 bushels was $10,000 to $12,000 in the early 1930's (*The Canadian Grain Trade*, p. 93). Twenty years earlier the Saskatchewan Elevator Commission reported the cost of construction of a country elevator with a capacity of 25,000 or 30,000 bushels to be approximately $6,000 to $6,500. An examination of the auditors' reports of farmers' elevators indicated to the Commission that these elevators had varied in cost from $6,200 for one with a capacity of 25,000 bushels to $11,000 for one with a capacity of 60,000 bushels, and that costs of construction per bushel of capacity had ranged from 18 cents to 26⅔ cents. See *Report of the Elevator Commission of the Province of Saskatchewan* (Regina, 1910), pp. 42 ff.

[4]By the Manitoba Grain Act, *Statutes of Canada*, 63–64 Vic., c. 39 (1900).

[5]The Saskatchewan Elevator Commission stated that over 3½ million bushels of wheat from the crop of 1908 were handled over loading platforms in Saskatchewan and upwards of 15 million bushels in the three prairie provinces. "Loading platforms are, therefore," said the Commission, "real competitors of the elevators, and their competition has tended to the protection of the farmers who do not use them [as well as those who do]." The saving in elevator charges through the use

The standard elevator has great advantages over flat warehouses and loading platforms. By a simple modification of a basic mechanical technique, metal cups were attached to an endless belt which, operated vertically, provided an effective power-driven instrument for lifting grain and distributing it into any one of a cluster of tall tanks or bins. Texture and consistency of dry cereal kernels *en masse* are such that gravity will move them readily through a small opening or spout; the bins, filled from the top, can readily be emptied from the bottom. A composite working unit embodying an elevating device is known as a grain elevator, or, simply, an elevator.[6] An elevator's essential features in addition to the elevating "leg" are a variable number of bins, a weighing and unloading platform with a receiving hopper beneath, and a hopper weigher for loading cars. The structures themselves vary. The giant terminals at tidewater, at the head of freshwater navigation, or less frequently at interior points, are constructed with concrete bins which are often combined into integrated series with a capacity of several millions of bushels for a single operating unit, a "terminal elevator." Local elevators of wooden construction throughout may hold from thirty or forty thousand to one hundred thousand bushels.

Local elevators, either a single one, or two or three at near-by railway points, have frequently been operated by individual proprietors, but more commonly by corporate owners, who centrally control groups of elevators running into scores or hundreds. In the American terminology which came to the Canadian plains with American technology, such institutions are dubbed "line elevators," and their operators, "line elevator companies."[7] While farmer-owned, or "co-operative," elevators constitute

---

of loading platforms was recorded by the Commission at $17 per car, or 1¾ cents per bushel on approximately 1,000 bushels, then regarded as a typical carload of wheat. See *Report of the Elevator Commission of the Province of Saskatchewan*, p. 74. Calculations of the monetary saving due to the use of the loading platform obviously take no account of the additional costs borne by the platform shipper in terms of added labour, inconvenience, and uncertainty. The federal Royal Grain Inquiry Commission appointed in 1923 stated that loading platforms served a useful purpose and recommended that the statutory requirement for their establishment and maintenance by the railways should be continued. See *Report of the Royal Grain Inquiry Commission* (Ottawa, 1925), p. 35.

[6]Dr. MacGibbon states that the grain elevator made its first appearance in Buffalo in 1841, and that it spread from there to other terminal points such as Chicago. It was gradually adapted to the requirements of local grain buying centres and came into use throughout the wheat producing area. See MacGibbon, *The Canadian Grain Trade*, pp. 85 ff.

[7]One American authority states that the groups of elevators under common ownership came to be known as line elevators because they were usually located along a single line of railway. See H. C. Filley, *Cooperation in Agriculture* (New York, 1929), p. 30.

the largest individual groups of elevators in existence in the Canadian wheat economy, and have for many years, they have nevertheless never been described as "line" elevators. The latter term has, from the earliest days of western farm organization, been exclusively applied to elevators operated independently of farmers, that is, by "private" capital. Elevators in the Canadian West have, therefore, been of two clearly recognized kinds: line elevators, and co-operative or farmer-owned elevators. A number of the important lines of Canadian elevators, local and terminal, were established in the early days of western development by milling companies and thus represented integration by the millers of the sources of raw materials with the processing operations.

In the earlier decades of the century, the local grain handling agencies operated in various capacities. Three distinct functional agencies or persons could, in fact, be recognized in local markets: the merchant, the warehouseman, and the agent or commission man. The merchant was the person, individual or corporate, who bought grain outright, ordinarily with the expectation of sale at a profit at a later time and, very commonly, at another place. He took title to the product and, apart from the possibility of some formal or informal offsetting or "hedging" transaction, assumed the risk of price changes throughout the interval during which title was retained. The warehouseman provided storage facilities and the equipment for weighing and transferring the product, the transfer ordinarily being from one transportation medium to another, from farm vehicle to box car. The agent or commission man acted for the seller or for the buyer; he did not take title or assume risk of price changes, but arranged for a sale or for a purchase on behalf of his principal. His payment was in the form of a commission which in the grain trade was customarily on a specific rather than an *ad valorem* basis.

These three functional agencies might be distinct and separate or might all be embodied in a single person or corporation. Independent grain buyers, grain merchants, operated at country points in the early days in considerable numbers without elevators of their own. They took out "space licences" and operated by purchasing wagon-load lots of grain offered for sale on the street. They leased special bin space in local elevators.[8] Agents or commission men frequently did business as track buyers in addition to conducting their commission business, each type

[8]For numbers, and the decline in numbers, of space licences from 1902–3 to 1909–10 see *Report of the Elevator Commission of the Province of Saskatchewan*, Table 12, p. 113. In the terminal field, "public" terminal elevator operators were licensed to perform but one function, that of warehousing. "Private," or "hospital" elevator operators, on the other hand, were licensed to act as merchants as well as warehousemen.

of activity, however, having to be separately licensed. Country elevators frequently served in the triple capacity of grain merchant, warehouseman, and commission dealer. Apart from the years when the Canadian Wheat Board has had a monopoly of Canadian wheat marketing (the crop year 1919–20, and from 1943 to the present time), country elevators practically always performed at least two functions, those of warehouseman and grain merchant. This association of functions is so thoroughly established in the grain trade that it might be assumed to be inevitable, but this is by no means the case.

Before 1943, country elevators in the prairie provinces entered the agricultural market as buyers of farm produce and as sellers of services, the services of warehousing and commission dealing. Grain growers, on the other hand, were in the market as sellers of farm produce and as buyers of the services associated with the marketing of their cereal products. Inequality of bargaining power in local grain markets has therefore had significance for two sets of prices: where the elevator companies acted as merchants to buy grain from farmers, the price at which the farmers sold their product was the important one; where the elevator companies acted as warehousemen or commission dealers and were, in effect, selling the services of such agencies to grain growers, there was the question of the price which farmers had to pay for such services. The distinction between the two sets of prices may have had no great practical significance, for inequality in bargaining power would apply equally to both. The distinction is nevertheless worth making in the interest of clarity. Western agrarian protest succeeded in bringing the prices for warehousing services and commission dealings under effective governmental control at a comparatively early date in the development of the wheat economy,[9] whereas the establishment of any satisfactory measure of control over the local price of wheat in relation to terminal or "spot" prices proved to be a much more difficult task.

Price has undoubtedly been the most important single concern of western wheat producers in their local market dealings but it has by no

[9]Under the Canada Grain Act, the Board of Grain Commissioners has long had authority to fix the maximum charges to be made for the discharging of grain into and out of elevators, for insurance against fire, for storage, cleaning, and the treatment and handling of grain while in any elevator. Regulations of the Board as applied to licensed country elevators have set maximum charges at a single figure for each kind of grain for the collective services of "receiving, elevating, spouting, storing and insurance against fire for the first fifteen days, and delivering into railway cars or other conveyances." See, e.g., Board of Grain Commissioners for Canada, regulation no. 21, effective August 1, 1949, published as Appendix no. 1 to *Hand Book on the Sale and Handling of Grain through a Country Elevator* (issued by the Board of Grain Commissioners for Canada: n.d., n.p.).

means been the only one. This is readily understandable since prices are monetary sums per unit of the product and vary with quality. Cereal prices are quoted in cents, or dollars and cents, per bushel, by weight, and vary according to grade. A further variation relates to the existence of different quantities of foreign materials including weed seeds, other varieties of grain, and broken or small kernels in the product. This variation is adjusted for by an allowance or reduction from gross weight known as dockage.[10] Four variables, therefore, combine to determine the cash yield of a grower's wheat: (1) the number of bushels gross weight; (2) dockage and shrinkage; (3) grade; and (4) price per bushel. Every one of these factors has been the cause of dissatisfaction to the growers in the marketing of western grain. The persistent agrarian belief in the early decades of the present century was that inequality of bargaining power in grain markets, associated with monopoly of line elevators, lay at the root of injustices in the determination of these items. This belief prompted insistent demands from grain growers for investigation and governmental control, and inspired the self-help efforts made possible by co-operative organization and the creation of co-operatively owned facilities.

Until 1943, Canadian wheat growers could deliver wheat to a local elevator under any one of a number of alternative arrangements. The basic choice was between delivery for storage and shipment, and delivery for immediate sale. There were, in turn, modifications of each of these arrangements. In delivering for storage and shipment the farmer had three options: (a) to take a graded storage receipt; (b) to take a graded storage receipt subject to inspector's grade and dockage; and (c) to deliver into a special bin. Delivery for immediate sale might take place on the basis of either the "street" or the "track" price. Wheat available in less-than-carload quantities is "street" wheat and must be disposed of on the basis of "street" prices. "Track" wheat is wheat loaded in box cars at local assembly points. The significance of these choices and options can be demonstrated in relation to the determinants of the value of a particular consignment of wheat, namely, gross weight, dockage, grade, and price.

[10]Shrinkage is a further deduction allowed from the gross weight of a farmer's delivery of grain to a local elevator. Shrinkage is defined by the Board of Grain Commissioners as "the invisible waste that takes place in the receiving, storing and shipping of grain." The maximum amount of shrinkage which may be deducted is not left to the discretion of the elevator operator but is specified in a table prepared by the Board of Grain Commissioners and required to be posted conspicuously in the local elevator. The shrinkage allowance is stated in pounds and varies only with the weight of the consignment and with the kind and condition of the grain. See *ibid.*

The significance of the distinction between street and track prices must be examined more fully later in this chapter. It can be pointed out here, however, that local sale on either the street or the track basis required agreement between buyer and seller on all points, namely, weight, dockage, grade, and price. The sale once effected, none of these factors could remain in dispute. In situations where delivery was for storage and shipment, however, the matter of price was not involved, but weight, dockage, and grade were. Acceptance of a graded storage receipt by the grower signified that agreement had been reached on weight, dockage, and grade. If agreement could not be reached on grade and dockage, and if a special bin was not available, the alternative was to accept a graded storage receipt subject to inspector's grade and dockage. Under this option the elevator operator drew samples from each load delivered by the grower, mixed them, and forwarded the blended sample to the nearest Dominion inspection office for a decision on grade and dockage. Storage under the third option, that of special binning, maintained the identity of the grower's shipment in a special bin. The storage receipt issued by the elevator operator under this option specified nothing but the weight of the grain delivered. The grade and dockage of the grain were determined by official inspection as it passed through an inspection point on its way to the terminal market. The elevator operator nevertheless preserved a sample of the grain which went into a special bin. If the grower was not satisfied with the inspector's grade, the sample could then be forwarded to the inspection office to determine, not whether the grade was correct, but whether the identity of the grain had in fact been preserved.

The obvious advantages of special binning from the grower's point of view would suggest that most grain going through storage in country elevators over the years must have been handled on this basis. This might well have been the case had it been only a matter of the grower's choice. It is, however, comparatively costly for an elevator company to provide special bins, because they waste space, and they have therefore been offered as a privilege rather than as a right. Elevator companies have used such facilities as a competitive device, charging only as much for their use, when granted, as for the use of storage facilities under either of the other storage options.[11]

Grain is stored in a country elevator in anticipation of sale. Before

[11]Regulation no. 21 of the Board of Grain Commissioners for Canada, as cited in n. 9 above, prescribed a uniform maximum charge for each kind of grain regardless of whether it was received as special binned grain or not. Board regulations of earlier years permitted a differential for special binned grain.

1943 it might be sold to the elevator company in whose local bins it was stored. An extremely important question over the years, however, was whether it *had to be* sold locally or whether it had an alternative outlet. It can readily be imagined that a grower who was forced to surrender title to his product at the local assembly point would, generally speaking, be in a much weaker bargaining position than if he had the choice of sending his grain to be disposed of in the terminal market. This distinction has, in fact, provided the basis for some of the most severe and most persistent of the western grain growers' criticisms of the line elevator companies and of the private grain trade generally. No attempt to modify the apparent inequity of the situation achieved appreciable success until co-operative elevator companies had been established by the growers and had spread throughout the wheat growing area sufficiently to offer effective competition to the line elevator interests.

The critical question was whether or not the individual grower was able to provide a carload of grain of a particular variety and grade. Because grain is handled in bulk and there is no bulkheading of cars, it has been imperative that western grain should move from the local assembly points to the terminals in carload quantities in order not to aggravate a freight burden already onerous. Minimum carload rates, when applied to a shipment smaller than that allowed for in the minimum carload calculations, rendered any appreciable undershipment impracticable. Grain cars have been increased greatly in size since the first decades of the present century when "small" grain cars had a minimum capacity of 1,000 bushels for freight calculation purposes and large cars had a rating of 1,400 bushels. Modern all-steel grain cars have a rated capacity as high as 2,000 bushels.

Regarding the situation superficially and having in mind the ordinary railway-poster publicity which has made of every western homestead a teeming bonanza farm from the earliest days of western settlement, it may be difficult to imagine how the problem of filling a one-thousand bushel box car with wheat could give rise to serious difficulty. Nevertheless the difficulty was acute and prolonged and gave rise to one of the major frictions which characterized western farmers' market dealings until well into the present century.

The difficulty caused by the indivisibility of the box car, particularly in the early years or decades of western pioneer life, is understandable in view of all the circumstances. The frontier homestead comprised 160 acres of unbroken sod. Pre-emption rights, after their restoration in 1908, or the purchase of an additional quarter-section from the Canadian Pacific Railway Company or the Hudson's Bay Company increased many of the original homesteads to 320-acre holdings. While many larger farms emerged, and while the breaking of sod on some farms of all sizes was accomplished quickly by the use of multiple gang-ploughs drawn by steam engines or "oil-pulls," the great majority of homestead holdings were broken slowly and with tedious labour by ox- or horse-drawn breaking-ploughs. Summer fallowing of one-third of improved acreage was common practice in the western wheat economy from pioneering days onward as a method of dry-land farming. For a variety of reasons, therefore, the annual acreage in wheat or in any other particular cereal was comparatively small on a great proportion of farms.

Other factors as well impeded the carload shipment of grain by the individual farmer. Crop yields were by no means always good. The output of a particular variety of grain from a particular farm was not necessarily all of one grade in any one year. Not infrequently the grain derived from a single field was accorded a progressively diminishing grade because of bleaching or increase in moisture content caused by rain or snow which intervened between the commencement and the completion of threshing. The problem of transporting grain from the farm to the local market centre in large quantities in a limited time frequently precluded the loading of carloads of grain over the loading platform and even rendered it difficult to assemble a carload in the local elevator within the fifteen-day period of storage covered by the minimum handling charge. The fact that a high proportion of western grains could not move to terminal markets in neat, single-grade carload lots is not surprising. In the middle twenties over 50 per cent of all western grain was still bought by country elevators in less-than-carload lots.[12] In earlier decades the percentage had necessarily been larger.

The significance of these facts requires little demonstration. A grower with one or more carload quantities of wheat or other grain in store in a country elevator might, of course, sell it locally to the elevator company. If, however, the elevator company offered a price considered by the grower to be out of line with terminal prices as indicated by daily quotations, the grower could instruct the elevator company to load his grain and forward it for sale in the terminal market. Under these circumstances the local price for grain in carload lots could never be depressed far below terminal prices, all costs considered, except in a situation marked by acute and persistent congestion of transportation facilities. For quantities of grain, which for any reason were less than sufficient to fill a box car at least close to its minimum capacity, there was no alternative but to sell it to the elevator operator, who could then make up car-

[12]MacGibbon, *The Canadian Grain Trade*, p. 102.

load lots by combining the deliveries of the same grade from several farmers. The competitive bidding of buyers in the terminal market had no direct effect on the local price of such grain. The spread between terminal prices and the local price of less-than-carload lot grain would be narrowed to a cost-plus-reasonable-profit basis only if there was, in fact, genuinely competitive bidding among elevator companies in any particular local market centre.

The significant economic characteristics of track wheat are, first, that it is in carload lots and, second, that it is in box cars. The latter point may appear to be merely a statement of the obvious, but it has been of considerable importance, not because loading wheat into a railway car is a difficult or costly process, but because, during the marketing rush in the fall of the year, securing a railway car to load has frequently been a matter of the utmost difficulty. Barring a congestion in transportation, however, the fundamental economic characteristic of track grain is that it is available in carload quantities and thus can move readily to the terminal market. It has the quality of mobility which is of such fundamental importance in terms of competitive competence and economic health. Street wheat or grain, on the other hand, lacks that important quality. It does not exist in carload lots while owned by individual farmers. Before 1943, therefore, it had to be sold to the local buyer and, generally speaking, on his terms regarding price, weight, grade, and dockage, if it was to be disposed of at all.[13] In local market centres where there was adequate competition among elevator companies, no problem existed. Where competition did not exist or was limited in extent, the conditions of sale were likely to be highly unsatisfactory to the grower.

Emphasis has been placed earlier in this analysis on the capital-cost burden entailed by the establishment of an adequate elevator system for the wheat economy. Railways constructed throughout the wheat growing region could derive earnings from grain movement only if initial and terminal transfer facilities were provided to make such movement not only possible but economical. Persons acquainted with the history of the financial struggles of the Canadian Pacific Railway Company in its early

[13]Co-operative elevator companies and the wheat pools in western Canada have always recognized street wheat as distress wheat and have followed a deliberate policy of narrowing the spread between street and track quotations. Toward the end of the period of contract pooling of wheat, Dr. MacGibbon reported that the spread between track and street prices at local points was 5 or 6 cents per bushel and that the net advantage to the track seller, after adjustment for hidden costs, was estimated at 3 or 4 cents per bushel. See *ibid.*, pp. 102–3. Dr. MacGibbon's book was published in 1932, its Preface is dated January 15, 1932, and the writing may be presumed to have been largely completed before mid-summer, 1931, while contract pooling was still in operation.

years will not be surprised to note that it did not set out to equip its many local sidings in the West with elevator facilities. The Company did construct a terminal at Port Arthur in 1883–4, the first one in the West, and others in later years, but country elevator facilities were left to be provided by other interests. The Ogilvie Milling Company constructed the first standard elevator in the West, at Gretna, Manitoba, in 1881.[14] but the extension of similar equipment lagged seriously behind the extension of settlement.

Flat warehouses and loading platforms were poor mechanical substitutes for standard elevators, but, paradoxically, the rapid construction of elevators throughout the West immediately prior to the turn of the century was accompanied by an intensification of dissatisfaction with grain handling facilities and by an accumulation of distrust of their operators. Western spokesmen in Parliament described the elevator system in the late nineties as "enlarged and fairly complete in Manitoba and the North-West Territories."[15] But this enlargement of facilities, far from assuring adequate and convenient transfer facilities to the growers at competitive costs, served only to convert local grain buying and the local sale of storage and grain handling facilities into a market situation in which monopolistic elements appeared to dominate. To western growers the destruction in the 1890's of the competition which had formerly existed at local buying points was associated with three circumstances: the extension through the West of a high degree of concentration of ownership of elevator lines; the establishment of the mechanism for the maintenance of price uniformity among the powerful elevator lines; and the actions taken by the Canadian Pacific Railway Company in its efforts to encourage the construction of standard elevators along its lines which ran through the wheat growing region.

R. L. Richardson, editor of the Winnipeg *Tribune*, and Liberal member for Lisgar, put the matter before the House of Commons in 1898:

These elevators [in Manitoba and the Northwest Territories] are controlled largely by syndicates—the Northern Elevator Company, the Manitoba Elevator Company, and others. In addition to these, there are the elevators owned by the Ogilvie Milling Company, and the Lake of the Woods Milling Company. So long as there was no combination, there was real competition in the buying of wheat. But these various syndicates put their heads together last season [1897], and, seeing that the price of wheat was considerably advanced, they decided to corral a larger share of the profit than was their legitimate due. So they formed what is popularly known in that country as

14*Ibid.*, p. 86.
15R. L. Richardson (Lib.) in Canada, *House of Commons Debates*, 1898, p. 2065.

a "syndicate of syndicates." Instead of each syndicate sending out a bulletin to their buyers at the various points throughout the province and territories, stating the price at which they wanted them to buy, the "syndicate of syndicates" meets in its little room in Winnipeg, each morning, and decides what price they propose, in their majesty, to allow the farmer for his wheat. Instead of various telegrams being sent out, the "syndicate of syndicates" sends out one telegram to each important point, and the buyer who receives it goes to the others, and the result is, that no one buyer at any station will pay more than the price which has been decided upon by this clique in Winnipeg.[16]

The practice outlined here foreshadows the formation of the North West Elevator Association, and eventually the North West Grain Dealers' Association in 1903, and indicates their functional significance as "open price" associations.

On the day that Richardson outlined his views, and immediately preceding him, James Douglas, member for Assiniboia East, spoke to the introduction of a bill (no. 19) to regulate the shipment of grain in Manitoba and the Northwest Territories. His emphasis also was on the emergence of monopoly. He pointed out that grain buyers who had formerly operated through flat warehouses had for the most part changed to elevator operation on the basis of an agreement with the Canadian Pacific Railway Company that flat warehouses should no longer be used. There were now, he said,

a number of elevators in full operation, and an understanding has been reached between the Canadian Pacific Railway and the elevator syndicate that all such flat warehouses must pass out of existence; so that in the province of Manitoba and a portion of the North-West Territories, the business has been wholly absorbed by the elevator syndicate. . . . It is only during the past season [1897] that this arrangement has been between the Canadian Pacific Railway and the owners of the elevators. . . . Under this agreement the producer is absolutely in the hands of this combine which is known to exist. It is, if I may say, a syndicate of syndicates which has formed an agreement with the Canadian Pacific Railway. The agreement appears to be this, that where an individual proposes to put his capital into the building of an elevator at any shipping point, the Canadian Pacific Railway will protect the capital from being affected by the building of another warehouse at that particular point.[17]

[16]Ibid. Richardson was speaking to the introduction of an amendment to the Railway Act "which would compel the companies to provide cars at the various points." He said, "Now we may not be able to regulate combinesters, but this House is able to regulate the railway companies. . . ." The situation in regard to monopoly in the elevator business was, he said, much worse in 1897 than it had formerly been. In earlier years there had been active competition at "nearly all the important market points," the competition provided by as many as five or six grain buyers at each market centre. Ibid., pp. 2066–7.

[17]Ibid., pp. 2061–2.

Douglas went on to point out that at Fleming, Moosomin, and a number of other points, warehouse operators had been notified by the railway company to build standard elevators or to close up. He gave the estimate of the Premier of Manitoba that "this agreement between the Canadian Pacific Railway and the elevator syndicate amounts to no less than one million dollars [taken from the farmers of Manitoba] for the past season."[18]

In 1899 the Dominion government appointed the Royal Commission on the Shipment and Transportation of Grain in response to allegations such as those of Douglas and Richardson. The Commission was instructed to investigate charges of unfair dockage and false weights at local grain handling points in the West, and the charge "that the owners of elevators enjoy a monopoly in the purchase of grain by refusing to permit of the erection of flat warehouses where standard elevators are situated, and are able to keep the price of grain below its true market value to their own benefit and the disadvantage of others who are specially interested in the grain trade, and of the public generally."[19] With particular reference to the latter point the Commission reported its findings as follows:

There were, and are, at most shipping points more than one elevator, so that a farmer could generally choose to which he would sell. The evidence, however, shows that *in many cases there is little, if any, competition between elevators as to prices* and that there is seldom any advance from other buyers on the offer made to a farmer by the first buyer he approaches. Of late years there have been combinations of elevator owners into large companies. This has resulted in fewer and larger elevator-owning corporations, which naturally tends to further decrease competition.

The evidence shows that there are now 447 elevators (exclusive of terminals) in the Manitoba (Western) inspection district, owned as follows:

Three line elevator companies own	206 elevators
The Lake of the Woods Milling Company own	50
The Ogilvie Milling Company own	45
Farmers' Elevator Companies own	26
Individual millers and grain dealers own	120
Total	447 [20]

By the beginning of the new century, disparities of bargaining power in the markets in which western farmers disposed of their produce served as the focus of agrarian protest in western Canada. As the preceding

[18]*Ibid.*, p. 2060.
[19]See *Report and Evidence of the Royal Commission on the Shipment and Transportation of Grain*, 1900, Canada, *Sessional Papers*, 1900, nos. 81–81a, p. 1.
[20]*Ibid.*, p. 9. Italics added.

references make clear, these disparities were attributed by the farmers to a deliberate and increasing curtailment of competitive action among grain buyers and warehousemen at local market centres. The Manitoba Grain Act, passed by the Dominion government in 1900, provided for a thorough system of regulation of the grain trade but, beyond new regulations concerning the provision of loading platforms and the allocation of cars, it contributed little toward the restoration of competitive balance in cereal markets. It was, moreover, clear from the start that the activities of local elevator operators and of most grain buyers were controlled by the head offices of their respective companies in distant places. These offices were centrally located in Winnipeg, in the "terminal" market. In 1887 the grain trading interests organized an exchange in Winnipeg and early in the new century made provision for futures trading. Four hundred miles beyond, at the head of the Lakes, were the great inland terminal elevators. Here again were the elevator companies, operating, for the most part, on an integrated basis to the extent that "line elevator" companies owned and operated terminals, or, looking at it the other way around, terminal elevator companies operated lines of country elevator houses.

By 1905 the protests of western grain growers were being formally expressed and put on record by their associations. The Royal Commission on the Grain Trade of Canada appointed by the Dominion government in 1906 in response to western complaints,[21] and in anticipation of a general election, was directed to "take into consideration all or any matters connected with the Grain Inspection Act and the Manitoba Grain Act. . . ."[22] The list of matters brought to the Commission's attention provides a fair indication of difficulties which were so persistent in the pre-1914 years that they may well be regarded as chronic at that time:

Among the matters brought to our attention [the commissioners reported] were: Improper weighing at country elevators; excessive dockage taken; returning of screenings to farmers at country elevators, and the allowance for screenings at terminals; special binning; car distribution; car shortage; complaints against the Inspection Department; the operation of elevators in Winnipeg, Fort William, Port Arthur and eastern ports; complaints con-

[21]In 1906 an agricultural deputation appeared before the Agricultural Committee of the House of Commons at Ottawa to protest the inadequacies of grading and inspection and other of the facilities for the marketing of western grain. H. S. Patton, *Grain Growers' Coöperation in Western Canada* (Cambridge, Mass., 1928), p. 42.

[22]See *Report of the Royal Commission on the Grain Trade of Canada, 1906,* Canada, *Sessional Papers,* 1908, no. 59, p. 3.

cerning the inspection of grain at Montreal; . . . complaints that prices paid in the country were too low, and the spread between track and street wheat was too great.[23]

The complaints include no specific reference to monopoly although it is clear that a number of the practices complained of could not exist under actively competitive conditions.

So many of the accusations of monopoly in the grain buying activities of the private grain trade have centred around the North West Grain Dealers' Association that it is worth noting in some detail the nature of this Association as determined by the Commission in 1906. The Commission found it to be a corporation chartered by the province of Manitoba, with a membership made up of elevator owners and grain dealers, many of whom were also members of the Winnipeg Grain Exchange. The Association had but two business activities: buying supplies for the country elevators of its members; and sending joint telegrams containing information on prices to country buying points where any of its members had elevator houses. "The main object of the association evidently," the Commission reported, "is the regulating of the buying of grain in the country."[24] The daily telegram gave information concerning both track and street prices. While the Commission found that the daily track price was fixed absolutely by the closing cash price in the Winnipeg Grain Exchange, the street price, it found, "is variable and depends upon the car supply, the price of the delivery month in which the grain may reasonably be expected to arrive at the delivery point, and *such expenses and profits as are agreed upon* from time to time by the members of the association."[25] Summing up its analysis of the Association and its activities, the Commission said:

While it is quite evident that there is an agreement or understanding that the list prices as sent to the country and the changes as wired from time to time are to be adhered to and buyers instructed to this effect, we cannot find anything in the association by-laws compelling its members to abide by the prices so decided upon, nor can we find any penalty provided for the breaking of the prices. We find that these prices are not adhered to in all cases, although *where a buyer persists in breaking prices he is brought into line by the combined action of other buyers on prices.* . . . This system of the North West Grain Dealers' Association is no doubt a trade restriction, but whether it

[23]*Ibid.,* p. 5.
[24]*Ibid.,* p. 16.
[25]*Ibid.* Italics added. As Dr. MacGibbon explained the procedure many years later, the telegrams sent out by the North West Grain Dealers' Association gave *actual* track prices and *changes* in street prices. See *The Canadian Grain Trade,* p. 313.

constitutes an undue restriction or not is a matter which is now before the courts.[26]

The reference which the Commission made to the courts will be elaborated below. At this point it is appropriate to refer to one other activity which the Commission encountered in its examination of the grain trade, an activity clearly directed toward the curtailment of competition in the local grain buying business. The Commission reported that there "had been" an agreement among certain elevator companies to pool their receipts or earnings at the various points where their elevators came into competition. Each elevator was allocated a certain proportion of the total grain receipts at a buying point. Adjustments were made following the presentation of monthly reports: those companies that had received more than their allocated percentage paid into the pool; those that had received less, drew from it. "This pooling arrangement," the Commission remarked, ". . . placed such a restraint upon the operations of the elevator companies within the agreement that it constituted a menace to those who had to sell grain to these elevators, and tended to unduly limit competition."[27] Although it found that the pooling arrangement had been discontinued, it recommended that such activities should be made illegal in Canada as they had been in Minnesota.

Following the hearings of the Commission in Winnipeg in November, 1906, the President of the Manitoba Grain Growers' Association brought formal charges against three members of the Council of the Winnipeg Grain Exchange who were also members of the North West Grain Dealers' Association. The charges were that the companies represented by these persons had "unlawfully conspired, combined or arranged with each other, to restrain or injure trade or commerce in relation to grain." Preliminary hearings in January, 1907, resulted in an indictment on thirteen points against the defendants and the case went forward for Crown prosecution in the Assize Court. At the trial in April, 1907, the prosecutor sought to establish that the various activities of the defendants, including particularly the pooling agreement and the concerted fixing of prices at country points through the North West Grain Dealers' Association, were in violation of section 498 of the Criminal Code. The defendant companies argued that the pooling arrangements which had been in effect in 1905 and 1906 had been abandoned. As for the price quotations circulated by the Association, the defence was that the daily telegrams were merely indicative of the current prices, that there was no

[26]*Report of the Royal Commission on the Grain Trade of Canada, 1906,* p. 16. Italics added.
[27]*Ibid.,* p. 17.

agreement that they be adhered to, and that no penalty was provided for failure to adhere to the prices as quoted. Mr. Justice Phippen found that there was no "undue restraint" in the actions of the defendants and, furthermore, that "the very acts complained of, taken in connection with their surrounding conditions, made on the whole for a more stable market at the fullest values and so for the public good."[28] The Manitoba Appeal Court upheld the verdict.

The protests and investigations relating to the Canadian grain trade in the years 1906–7, coupled with the court actions of the same years, make clear certain central features of the western farmers' marketing situation in the early years of the century. By that time the North West Grain Dealers' Association was firmly established in practice[29] and, after 1907, fully sanctioned by the approval of the courts. This Association was a trade association of a particular type, an "open price" association. The daily telegraphic circulation of uniform price quotations by its secretary carried no formal commitment of adherence by the membership to the quoted prices and, certainly, no graduated financial penalties for departure therefrom. Nevertheless, from observing the activities of the open price associations of later decades, particularly those in the United States after 1920,[30] there can be no doubt of the validity of the wheat growers' claim that the intention and the certain if immeasurable effect of this Association after 1900 was to lessen competition in the local markets for Canadian grain. The establishment of the North West Grain Dealers' Association meant that western grain growers, who of necessity brought their grain to market under conditions of perfect sellers' competition, were compelled to deal with elevator companies whose competition among themselves as buyers was significantly curtailed by two factors: by a high degree of concentration of ownership and control of elevator facilities; and by the establishment of an open price association, a mechanism working strongly toward the maintenance of uniformity in price bids.

The Commission appointed in 1906 looked briefly at the problem of street grain (which, as already noted, created the chief price problem in local markets), but completely missed the significant points concern-

---

[28]Patton, *Grain Growers' Coöperation in Western Canada,* pp. 57–9, citing 6 Western Law Reporter 19 (1907).

[29]While the North West Grain Dealers' Association was not incorporated until 1904, there was nevertheless clear indication of its price-influencing activities at least as early as 1897. Cf. the speech of R. L. Richardson in the House of Commons in March, 1898, as cited in n. 16 above.

[30]See, e.g., Henry R. Seager and Charles A. Gulick, Jr., *Trust and Corporation Problems* (New York, 1929), particularly chap. xvi.

ing its economic position. "We have given the matter of street grain our serious consideration," the Commission said, "and we fail to agree as to whether or not the prices paid on the street are reasonable. The main protection . . . against undue depression is the right to ship grain to the central market and sell on a track basis. This protection will not be complete until such time as the car supply during the shipping season is ample for requirements."[31] Car shortages certainly complicated the problem, but even with adequate railway facilities and services the problem of street wheat remained, for such wheat could be disposed of only by sale to the local elevator operator.

Two other important points are made apparent by the investigations and court actions of 1906–7. The first relates to the state of Canadian law and the opinion of the courts at that time concerning combinations in restraint of trade. The question that exercised the court in considering the indictment of members of the Winnipeg Grain Exchange and of the North West Grain Dealers' Association in 1907 was not whether they were operating as a combination, nor whether they were operating in restraint of trade, but only whether their activities constituted *undue* restraint of trade. The determination of how much restraint was necessary before it could be classed as undue lay with the courts, and the decisions of 1907 made it clear that the curtailment of competition in the Canadian grain trade would need to progress to considerably greater lengths than it then had before Canadian courts would hold it to be an encroachment of section 498 of the Criminal Code. The first Combines Investigation Act was enacted in the same year but, as pointed out in the previous chapter, neither the first and thoroughly useless Combines Act of 1907 nor the rather more efficacious one of 1923 altered in any way the basic Canadian approach to the problem of monopoly. Among its most important underlying assumptions were these: that monopoly was not inherently bad; that combination in restraint of trade and in curtailment of competition was not necessarily detrimental to the public good; and that, with rare exceptions, business men were reasonable and well intentioned, and their actions would therefore not likely be found to be unreasonably neglectful of the public interest.

The other point illustrated by the events of 1906–7 is that while western growers at that time brought vigorous complaint against individual members of the Winnipeg Grain Exchange they were, in fact, complaining against what appeared to them to be a misuse of that

[31]*Report of the Royal Commission on the Grain Trade of Canada, 1906,* pp. 16–17.

institution rather than against the institution as such. The curtailment of competition effected by Exchange members by means of the North West Grain Dealers' Association, and by certain of the usages and rules agreed upon in their capacity as Exchange members, was regarded by western growers as contributing to gross inequality of bargaining power between sellers and buyers in local markets. But these actions were not regarded as inherent in the operation of a grain exchange, and the agricultural protest against such actions did not rest upon nor imply dissatisfaction with an exchange and futures market if freed from anti-social, monopoly elements.

This distinction is important. It serves to explain the sharp differences that existed in the views of western agricultural leaders after 1920 as compared with their views before that date on the methods to be pursued in the marketing of western grain. The Commission of 1906 was made up exclusively of western farmers. They reported minor objections to one or two of the rules of the Grain Exchange.[32] In general, however, they gave the Exchange itself a clean bill of health. "The prices at which transactions are made [on the Exchange]," they said, "are officially posted on the blackboard by a man provided by the Exchange. These prices we find are made in open competition, and are beyond doubt the full value of the grain as based on the world's markets. The work of the Grain Exchange in establishing and systematizing a market in Winnipeg for the handling of the crops of the West has been a great benefit to the country. . . ."[33]

By the end of the first decade of the present century the conviction that monopoly elements dominated the grain trade had become so deeply and so generally rooted in the minds of western grain growers that they pressed strongly for the socialization of the elevator system at both its local and terminal levels. For many farmers there was every readiness to replace private monopoly with public monopoly; for others it would be sufficient to provide an adequate network of government-owned country and terminal elevators to offer an alternative to the private line and terminal elevator companies. The outcome of this agitation will be considered in the following chapter. It is mentioned here because its investigation by the government of Saskatchewan led to the preparation of a report which, among other things, gives the best

[32]The Commission felt, for example, that the selling commission fixed by the Grain Exchange should not be as high on barley and oats as on wheat, and that no by-law of the Exchange concerning such commission should be permitted to restrict competition among buyers at local points. *Ibid.*, pp. 14–16.

[33]*Ibid.*, p. 14.

available summary of the grain growers' pre-war complaints against the private grain trade.[34] Furthermore, the report presents these complaints in their proper economic setting as implying an all-pervasive monopoly in the purchasing of Canadian grain and in the provision of the elevator facilities for merchandising it.

The Saskatchewan Elevator Commission, the group which conducted the investigation referred to above, summarized in one chapter of its report the "charges against the present system" which were laid before it and which, as is abundantly clear from other evidence, were prevalent in the West in the pre-war years. These complaints were classified as being directed against the local and terminal elevators, against the millers, the grading system, the railways, the banks, and, lastly, against the Winnipeg Grain Exchange.[35]

Against the local elevators the allegations were of short weight, excessive dockage, improper grading, unduly low prices, refusal to provide special bins, the mixing and substitution of grain, and other minor malpractices. Informants were of two opinions on the responsibility for such actions. Some held that the actions were performed by the local elevator operator on the basis of instructions from the head office of his company. Others regarded them as the sole responsibility of the local operator. The belief was common that the local agent was **under such pressure** from his employer to make his elevator show a **profit that** various sharp practices were inevitable.

The chief complaints against the terminal elevators concerned excessive dockage, undercleaning, and mixing. The underlying belief in regard to the latter matters was that by undercleaning and mixing, the terminals diluted Canadian grain exports to the minimum quality in each grade while, on the average, farmers delivered grain of the *average* quality in each grade. Prices in overseas markets, it was argued, were depressed to conform with the inferior quality of overseas deliveries and farmers were accordingly paid low prices for grain of superior quality. The chief criticism against the western Canadian millers was that they picked the choice carloads of wheat for their own mills at going prices, skimmed off the top quality, and thereby contributed—as did mixing operations of the terminal elevators—toward a depreciation of the quality, reputation, and price of Canadian wheat in overseas markets. Working in the same general direction, it was argued, a grading system which stressed such functionally irrelevant characteristics as colour and

[34]*Report of the Elevator Commission of the Province of Saskatchewan.*
[35]For discussion of Winnipeg Grain Exchange, see chap. x, pp. 179–95.

plumpness of kernel contrived to favour millers and exporters at the expense of the farmers.

A great deal of the dissatisfaction with the railways had been disposed of by 1910. The monopoly clause of the Canadian Pacific Railway Company's charter had long been history, and by this time two additional transcontinental systems were spreading a network of lines throughout the West. After 1900 the Manitoba Grain Act had required the construction of loading platforms on the basis of local petition. Amendments to this act in 1902 and 1903, fortified by the Sintaluta test case of 1902, had forced the railways to institute and abide by the car order book. The railways were required by these amendments to supply each of their local agents with a standardized car order book. The agent, in turn, was required to enter all applications for grain cars in the book in numerical order and cars were to be distributed strictly in order of application. Each application, whether from a farmer or from an elevator agent, was to be for one car only, and no party was allowed to have more than one unfilled application in the book at any one time. Nevertheless, the protests against the railways in 1910 implied continuing distrust and irritation. The railway companies were accused of being generally obstructionist and disagreeable to deal with. They supplied leaky cars, it was stated, and "they construct loading platforms as if the object was to render the use of them by the farmers as difficult as possible."[36] They had lent assistance to the elevator monopolies in the past and they continued to favour the large milling and elevator companies at the expense of the farmer.

As for the banks, western farmers' criticisms of these institutions were legion and were dealt with by other investigating bodies.[37] One complaint in particular, however, was relevant to an examination of the grain trade and its short-comings. This was the complaint that the shortness of the term of the ordinary bank advance and the banks' insistence on an early fall clean-up of outstanding agricultural loans forced farmers to sell their grain immediately after threshing, thus placing the western crop in a disadvantageous position on a glutted market.

[36]*Report of the Elevator Commission of the Province of Saskatchewan*, p. 20.
[37]See, e.g., *Report of the Agricultural Credit Commission of the Province of Saskatchewan*, 1913 (Regina: Government Printer, 1913). An investigation into agricultural credit was carried on in Alberta a decade later. See *Report of the Commissioner on Banking and Credit with Respect to the Industry of Agriculture in the Province of Alberta*, D. A. MacGibbon, Commissioner (mimeo.; Edmonton, Nov. 4, 1922).

The Saskatchewan Elevator Commission regarded the local elevators as its special concern. It outlined a plan for a system of co-operative local elevators and urged that such a system be introduced in Saskatchewan. Its conclusions regarding local grain markets provide a concise summary of the influences at work and of the developments which had taken place in these markets by 1910:

The commission cannot believe that the increased railway facilities, the extended use of the loading platform, the work of The Grain Growers' Association and of The Grain Growers' Grain Company, the competition of the farmers' elevators, the introduction of public weigh scales, and the provisions of The Manitoba Grain Act have had no effect upon the initial elevators. They cannot believe that the excessive storage capacity has had no effect in stimulating competition. They cannot believe that companies would sell out elevators cheaply if they had in these sources of large profits. They are constrained to accept the testimony of many farmers to the effect that the conditions have been improved and that the man who knows can protect himself so far as the initial elevators are concerned.

The commission do not say that the conditions are always what they should be, that there are no cases of sharp practice, and that there are no grounds for such dissatisfaction as exists. They are impressed by the existence of a very strong feeling of dissatisfaction on the part of some farmers who cannot be regarded as incompetent in their business or as mischief makers or agitators. The commission believe that behind such feeling there are experiences of rank injustice, recollections of times when the elevator operators had the farmers in their power, and when they took full advantage of their opportunity. The commission believe that the elevator companies brought the trouble upon themselves in earlier days. But they believe also that the situation has been materially improved by the factors referred to. It appears to the commission that these factors can be so strengthened by the province that the result would be to give the farmer complete control in the matter of initial storage of the grain.[38]

[38]*Report of the Elevator Commission of the Province of Saskatchewan*, p. 82.

CHAPTER EIGHT

Co-operation and Provincial Elevator Policy

ONE POSSIBLE SOLUTION for a community where local grain handling facilities were inadequate was for the farmers to raise the necessary capital and construct and operate an elevator of their own. The minimum capital requirements for such a project were not large. The operation of a line comprising scores or hundreds of country elevators and a terminal or two was, of course, big business, and not the sort of activity to be readily contemplated by new settlers who were short of capital for the development of their own farms and who had not yet established a tradition of large-scale co-operative endeavour. But elevators could be operated on a multiple or a unit basis. The immediate requirements of the farmers surrounding any shipping centre were, after all, for but one efficient and trustworthy country elevator located at their shipping point. Such a house might be large or small, but the generally accepted capacity over the turn of the century varied from thirty to thirty-five thousand bushels.[1]

Under the pressure of dissatisfaction with the services rendered by "outside" grain dealers—by local elevators built and operated by line elevator companies or powerful milling concerns—western farmers in a number of places were able to raise sufficient capital for the construction and operation of individual, local, farmer-owned elevators. In 1899 there were twenty-six farmers' elevators in Manitoba and the Northwest Territories.[2] In 1910 there were sixty such elevators—twenty-six in Manitoba, twenty-nine in Saskatchewan, and the remaining five in Alberta.[3] These were all operated on a unit basis, each at a

[1]The Canadian Pacific Railway Company provided free sites for the construction of "standard" elevators provided their individual capacity was at least 25,000 bushels.

[2]*Report and Evidence of the Royal Commission on the Shipment and Transportation of Grain*, 1900, Canada, *Sessional Papers*, 1900, no. 81a, pp. 7–9.

[3]*Report of the Elevator Commission of the Province of Saskatchewan* (Regina, 1910), pp. 88, 113–17, and *passim*. At the same time there were 1,830 country elevators in the prairie provinces, including the 60 farmers' elevators. At a number of points the farmers or the municipality had constructed and operated weigh scales. "And where the farmers weigh their grain before delivering it to the elevators," said the Commission, "the trouble about weighing has practically disappeared." *Ibid.*, p. 75.

separate shipping point and not under common ownership. As yet there was no farmer-owned terminal elevator in Canada. The Grain Growers' Grain Company, however, operated as a farmer-owned commission agency after its organization in 1906 and thus provided an important protective service in the merchandising of grain in the terminal market.

The Saskatchewan Elevator Commission of 1910 devoted consider-able attention to the experience of the farmers' country elevators with a view to advising the government on the formal request expressed by the Saskatchewan Grain Growers' Association for the establishment of a provincially owned and operated elevator system. The Commission found the record of these elevators spotty in the extreme. A number of them operated at a profit in particular years but the financial outcome of their operations was highly variable. Many had lost money every year. The Commission cited figures for the operations of a number of them which operated exclusively on a warehouse basis.[4] These elevators handled from one to two hundred thousand bushels of grain each in the year for which the data were presented. After allowing for interest and depreciation, the financial results of the individual elevators varied from a net loss of $465 to a net profit of $830. The figures for the salary of manager ranged from $500 (plus a $75 "donation") to $830 and were regarded by the Commission as exceptionally low. In attempting to ascertain the annual cost of operating a country elevator the Commission found a considerable range of estimates and experiences. Advocates of a provincial elevator system estimated the cost at $1,200,[5] but none of the records of the operating units confirmed this as a reasonable figure. Operating costs, interest and depreciation included, ranged rather from $1,850 to $2,750 and, on a comparable basis, might be said to average $2,400.[6]

In order to ascertain the sources of income available to country ele-vators, it is necessary to recall that their operators, in the early decades of this century, might act as warehousemen exclusively, or as warehouse-men and merchants combined. An elevator operator acting exclusively as a warehouseman could secure income only from rendering the ser-vices of receiving, storing, and loading grain into cars. The customary

---

[4]That is, they did not act as merchants, buying and selling grain, but merely accepted grain for transfer and storage. *Ibid.*, pp. 47–51.

[5]*Ibid.*, p. 43.

[6]Dr. MacGibbon estimated the cost of operating a country elevator of 30,000 bushels' capacity in the 1920's to be from $2,500 to $3,250 exclusive of interest on investment, depreciation charges, and a share of head office expenses. He estimated the total of all items at $4,000. See D. A. MacGibbon, *The Canadian Grain Trade* (Toronto, 1932), p. 105.

charge for these services until 1921 was one and three-quarter cents a bushel. Storage without extra charge was limited to fifteen days, and additional storage was customarily provided at three-quarters of a cent a bushel a month. But the elevator operator might also obtain a licence as a grain merchant which entitled him to buy and sell grain. This provided the possibility of additional revenue from the sale of grain in terminal markets at an advance over the purchase price at the local point plus the expenses of forwarding it to the terminal market. As noted in the previous chapter, the chief possibility for gain arose from the purchase of street rather than track grain, that is, grain purchased "on the street" in wagon-load or less-than-carload quantities. With the opportunity for profit from such transactions there also lay the danger of loss, but a given margin could be protected with reasonable certainty under normal conditions by hedging operations.

Line and mill elevator companies were ordinarily licensed as merchants as well as warehousemen and conducted their country elevator business accordingly. Approximately three-quarters of the farmers' elevators examined in 1910 did the same.[7] It is clear that warehousing operations would not cover elevator expenses unless the turnover were substantial. It was occasionally stated as a rule of thumb that a country elevator would have to be filled three times a year to pay its way,[8] that is, an elevator with a capacity of thirty thousand bushels would have to handle upwards of one hundred thousand bushels of grain a year to pay. In addition to the income from the initial handling charge, there would be an indeterminate revenue from grain left in storage beyond the fifteen-day minimum. The Saskatchewan Elevator Commission calculated that the annual turnover necessary to permit a country elevator to pay its way on a warehousing basis was closer to four times than three.[9]

The importance of these turnover ratios becomes apparent from data indicating how completely unattainable they were on the average. In the prairie provinces by the early years of the present century, the capacity of country elevators was such that the grain marketings of a typical year would not fill them more than twice. For the crop year 1908–9 their capacity was approximately forty-three million bushels.[10] The total quantity of western grains—wheat, oats, barley, and flax—marketed through country elevators during the same year was eighty-two million bushels; elevators in Manitoba and Saskatchewan were filled

[7]*Report of the Elevator Commission of the Province of Saskatchewan*, p. 88.
[8]H. S. Patton, *Grain Growers' Coöperation in Western Canada* (Cambridge, Mass., 1928), pp. 15–16.
[9]*Report of the Elevator Commission of the Province of Saskatchewan*, pp. 45–6.
[10]See *ibid.*, Appendix, Table 4, p. 108, and Table 8, p. 110.

1.8 times on the average and those in Alberta 2.7 times. Whatever may have been wrong with the western elevator system at that time, it could scarcely be alleged that private capital had failed to provide sufficient local capacity. That fact, however, did not prevent vigorous and persistent agrarian pressure for the establishment of provincially owned elevator systems even if new elevators had to be constructed at every shipping point where old elevators could not be obtained. Nor did it prevent the co-operative elevator companies which were formed or which went into the elevator business after 1912 from establishing three elaborate lines of country elevators in the prairie provinces, most of them newly built rather than purchased.

The farmers' elevators in operation in 1910 fared better than the average in terms of patronage. During the first nine months of the 1909–10 crop year, the 1,830 country elevators in western Canada handled seventy-five million bushels of wheat, or an average of forty-one thousand bushels per elevator.[11] During the same period the 60 farmers' elevators handled an average of over fifty-five thousand bushels each. They were far surpassed, however, by the 83 elevators in the chain or line of the Lake of the Woods Milling Company, which averaged approximately seventy-nine thousand bushels. The 300 mill elevators in the four mill chains in the West averaged fifty thousand bushels. While these figures represent deliveries for only nine months, they nevertheless give a fair indication of relative patronage. Since deliveries during the remaining three months would be comparatively small, the figures also indicate that western elevators with a capacity of fifty-three million bushels by 1909–10 would, on the average, fall considerably short of a complete second filling of wheat over the entire crop year in question.

Despite the better-than-average patronage of the farmers' elevators, and despite the fact that most of them purchased grain as well as storing and transferring it to cars, the Saskatchewan Elevator Commission could only characterize the financial experience of these local institutions as unsatisfactory. "That too many of them have been financial failures is unfortunately true," the Commission said, "and of this failure there are two general causes, each of which manifests itself in several ways. The two are bad management and competition."[12] In elaborating these causes, the Commission pointed out that bad management showed up in the following ways: (1) Too few shares were sold in each elevator before construction. Shares could be sold only with difficulty after an elevator was built and therefore too few farmers had a direct personal interest

[11]*Ibid.*, p. 88.
[12]*Ibid.*, p. 85.

in the success or failure of each venture. The heavy burden of interest on borrowed capital constituted a disproportionate fixed charge which reduced dividends on share capital and created dissatisfaction. (2) Excessive straining after economy in the matter of salaries and wages left the elevators in danger of dependence on incompetent or dishonest managers. The farmers who owned and patronized the elevators frequently distrusted the managers, and accounts were commonly badly kept. The outstanding individual successes among the farmers' elevators as a group were identified with good management and with management which was well paid and employed on a year-round basis. (3) Shortages resulted from taking too little dockage to cover shrinkage, from loss in transit, and from the terminal dockage. (4) In some cases shareholders felt that directors or other shareholders interfered unduly with the management of the elevator with the result that it came to be regarded as the instrument of the privileged few.

The competition faced by the farmers' elevators, so the Commission reported, was that of loading platforms and of other elevators. Many farmers used the loading platform even at local trading points where a farmers' elevator was in operation. This saved them approximately $17 per car (1,000 bushels at 1¾ cents per bushel), and besides, the Commission said, some farmers "distrust all elevators." It was reported to the Commission that shareholders and even some of the directors of farmers' elevators hauled their grain to "company" elevators. The latter directed their competition against the farmers' elevators in particular. "Unfortunately, however," the Commission added, "some of the farmers' elevators were so managed that they were not hard to beat."[13] Dissatisfaction on the part of farmers with the management of their own elevators diverted substantial amounts of grain to the line elevator companies. One or two of the farmers' companies had attempted to adopt the principle of exacting a penalty of 1 cent per bushel, paid to their own elevator, on all grain taken to a rival concern.[14] This principle had, however, apparently not proved practicable and had been dropped.

The problem of street grain bedevilled the individual farmers' elevators and contributed much to their financial embarrassment as well as to the dissatisfaction of the growers with their operations. Any elevator company could avoid the responsibility of dealing in street grain by operating exclusively under a warehousing licence, and approximately one-quarter

[13]*Ibid.*, p. 86.
[14]This attempt will be recognized as the Canadian counterpart of the Rockwell penalty clause developed and put into practice by a group of grain growers of Rockwell, Iowa, in 1889. See H. C. Filley, *Cooperation in Agriculture* (New York, 1929), pp. 50–1.

of the farmers' companies adopted this expedient. A number of them rented space to grain firms and either allowed their own managers to buy grain for these firms, the managers being paid by the firms for this service, or they allowed the firms renting the space to put buyers "on the street." In either circumstance the managers of the farmers' elevators were likely to come under the suspicion of being subject to the influence of the private grain buyers. In any case, an elevator operating exclusively on a warehousing basis could pay its way only if it could be assured of a local patronage approximately double that of the average western elevator.

Furthermore, from the standpoint of what the farmers required in the way of elevator service, street grain could not be ignored. The Warehouse Commissioner reported that during the 1908–9 crop year there was bought on street 55 per cent of all elevator receipts of western Canadian wheat, 78 per cent of the oats, 70 per cent of the barley, and 73 per cent of the flax.[15] While some of this grain was undoubtedly disposed of on street out of choice, as the quickest way of securing ready cash if for no other reason, the greatest part of it was certainly sold as street grain because it was available only in less-than-carload quantities and could be disposed of in no other way. Any elevator company or system which failed to deal satisfactorily with street grain could not claim to be getting to the root of agricultural dissatisfaction with existing grain handling facilities, a dissatisfaction which although exaggerated was nevertheless based on clearly and repeatedly demonstrated inadequacies.

In final comment on the experience of the farmer-owned local elevators in western Canada before 1910, the Saskatchewan Elevator Commission pointed out that profit and loss statements failed to record many relevant points. "It is a mistake to say," the Commission urged, "that as a class farmers' elevators have been a failure. They have not been a failure, in spite of all their difficulties."[16] By co-operative action in erecting and operating local elevators, the Commission said, the growers acquired first-hand knowledge of the elevator business and of co-operation. They protected themselves against injustice, and the competition which their operations afforded was of benefit to the whole farming community at the various points where farmers' elevators existed. "They have demonstrated," the Commission concluded, "that under certain conditions, which are not at all unattainable, farmers' elevators can succeed, and that co-operation as a solution of the question of the initial storage cannot be lightly thrown aside."[17]

[15]*Report of the Elevator Commission of the Province of Saskatchewan*, p. 87.
[16]*Ibid.*
[17]*Ibid.*

It is not commonly remembered that the elevator problem in western Canada was tackled by at least one operating scheme of municipal elevators.[18] At the beginning of the century, when the growers in many districts of the West were asking themselves what might be done to relieve them from their dependence on line and mill elevator companies, the farmers of the rural municipality of South Qu'Appelle were at least as exercised as any others. One proposal put forward in that community as in many others was that a joint-stock, farmer-owned company be organized for the construction and operation of elevators at the local loading points. The realistic argument was advanced, however, that in case such elevators were to incur operating losses the burden would fall exclusively on the farmers who had been sufficiently public-spirited to purchase shares and thus assist in the provision of capital. An alternative proposal was that the rural municipality should finance the construction of the necessary elevators by the sale of debentures, and that ownership and operation should be maintained as a public utility for the growers of the municipality. The assumptions were, apparently, that there was a task to be performed in providing effective competition against the line or mill elevators which were already operating locally, and that this task might well entail losses which should be borne by the ratepayers collectively and in proportion to their individual taxable assessments rather than by the farmer-ratepayers in the capacity of shareholders in the enterprise.

Reasoning along these lines, the Municipal Council of South Qu'Appelle undertook the establishment of two municipal elevators in 1902. The municipality sold twenty-year, 6 per cent debentures to raise $10,000 for construction. The two elevators, one at McLean and one at Qu'Appelle, each with a capacity of thirty thousand bushels, were built at a total cost of $10,260. Each year an "elevator tax" was levied upon the ratepayers of the municipality to meet the annual instalments of principal and interest due on the debentures outstanding, and to cover operating losses. The municipality operated the elevators, exclusively on a warehousing basis, for the five crop years from 1903–4 to 1907–8 inclusive. For the 1908 crop, both elevators were rented to private operators who paid a flat rental and undertook full responsibility for the operating costs for the season. In January, 1909, the McLean elevator was burned and the municipality recovered $3,750 in insurance. The Qu'Appelle elevator was sold for $3,000 in September, 1909, after the ratepayers had voted for the termination of the municipal venture on the grounds that there was no prospect of profitable operation.

[18]The information on municipal elevators is taken from *ibid.*, pp. 55–6, and Appendix, pp. 131–41.

Financially at least the municipal enterprise failed badly. The experience during the crop year 1907–8, which prompted the rental of the elevators to private operators, was particularly distressing. The crop was frozen and although the Qu'Appelle elevator handled fifty thousand bushels of grain the McLean elevator received only one thousand bushels which yielded the elevator an income of less than $20. Over the five-year period, the two elevators handled less than half a million bushels of grain between them. The largest volume handled by either of them in one crop year was 70,000 bushels. The Qu'Appelle elevator averaged 45,000 bushels a year and showed an operating loss in three of the five years; the McLean elevator averaged 40,000 bushels and lost money every year. The net operating loss for the total venture amounted to $1,600. Adding to this deficit a proper allowance for interest and depreciation, the total net loss to the municipality was approximately $8,650.

Although these elevators took grain only for transfer to railway cars and for storage, and purchased none, they nevertheless issued storage tickets indicating weight, grade, and dockage. They guaranteed grade and dockage as determined by their operators, and lost heavily in some years by being out of line with the official inspection. Their records suggest that they guaranteed weights as well. Because they operated on a warehousing basis, these elevators gave no direct assistance to growers whose grain was in less-than-carload quantities and had accordingly to be disposed of "on the street." They were as liberal as possible with special-binning privileges, a costly service which did not use their capacity to the best advantage, but this service also was available only for shippers of carload lots.

An auditor who investigated the records of the South Qu'Appelle municipal elevator scheme for the Saskatchewan Elevator Commission in 1910 interpreted its financial failure as due principally to the following causes: (1) losses from over-grading; (2) bad management; (3) lack of patronage; (4) failure to compete with the loading platform and with line and mill elevators situated at the same loading points; and (5) the establishment of loading and elevator points on new railway lines entering the district.[19]

The records of the municipal elevators showed that their operators apparently more than offset their losses from over-grading by the sale of grain secured as "surplus" by under-weighing. Bad management was particularly in evidence at McLean where "the several managers

[19]Ibid., p. 132.

appointed were not popular with the farmers."[20] The lack of patronage was closely related to this factor, but was influenced by crop conditions, by the establishment of new loading points within the district, and by competitive price-cutting on the part of local line elevator operators. It was reported to the investigator that the majority of the large farmers shipped exclusively over the loading platform, thus patronizing neither the municipal elevators nor their rivals.

The Saskatchewan Elevator Commission interpreted the municipal elevator experiment as throwing important light on the question of patronage. One of the common assumptions at the time of the Commission's investigations was that farmers could be counted on to patronize faithfully an elevator of their own provided it was a local enterprise. A further assumption was that even if an elevator system were organized on a large-scale basis by the government or by a farmers' co-operative, farmers would stand by the local unit of such a system simply because it could be regarded as more of a "farmers' " elevator than those belonging to line or mill elevator chains. The Commission's comment was:

The elevators would have paid their way had sufficient grain been taken to them. . . . Farmers [however] went where they believed they were getting the best terms. They had no personal interest in the municipal elevators. An individual's share of the [elevator] tax might easily be less than the amount gained by patronizing the other elevators. And even though the better terms in the other elevators might be due to the municipal elevators, the latter did not receive support enough to be able to survive. . . . local loyalty and local pride were absent.[21]

Locally owned and managed farmers' elevators were no match for the monopolistic elements which increasingly dominated the Canadian grain trade over the turn of the century. The powerful grain companies represented substantial aggregations of capital directed into the various areas of the grain marketing business on the basis of both integration and combination. The line elevator companies individually owned many country elevators, they were integrated in the field of terminal operation and the export business, and operated as commission agencies as well. The substantial chains of country elevators owned by the milling companies represented but one stage in the integrated operations of the wheat processing concerns. Much of the capital was American, and the Canadian organizations in certain important cases represented foreign

20Comment by A. W. Goldie, auditor and investigator for the Saskatchewan Elevator Commission. "The municipality," added Mr. Goldie, "has proved that everything depends on the capability and popularity of the men in charge." *Ibid.*
21*Ibid.*, p. 56. There were two other elevators at McLean and four others at Qu'Appelle.

investment of a direct or "branch plant" nature. Trade association of the open price variety further restricted whatever competition might otherwise have survived among the few dominant groups in the grain trade. Neither locally owned farmers' elevators nor independent private elevators could operate effectively within such an environment.

In the period ending in 1920 there were but two grain marketing ventures, or types of venture, which western grain growers put into operation on their own initiative. These were the individual farmer-owned and municipal country elevators described above, and the Grain Growers' Grain Company, a farmer-owned commission company operating in the terminal market. The latter proved to be an enterprise of the utmost importance in the improvement of the grain growers' position within the market, first at the terminal and later at local points. The enterprise was boldly conceived; it repeatedly appeared doomed to failure in its early years of operation, but, surviving, it demonstrated a remarkable soundness of functional design and this for two particular reasons. First, and unlike the projects for country or terminal elevators, it did not require heavy fixed capital equipment for its operations. Second, it operated where it could offer effective competition to private commission firms and whence it could readily extend its operations into the other areas of the marketing process. The details of its history are available in various sources,[22] but the main features may be given here. Of all the western co-operative enterprises, it, more than any other, was the creature of a single person. That person was E. A. Partridge. After a month spent in Winnipeg in the winter of 1904–5, delegated by the Sintaluta local of the Grain Growers' Association to investigate the conditions under which western grain passed through the Winnipeg inspection and marketing mechanism, Partridge returned home fully confirmed in a distrust of the private grain interests and their activities in and about the Winnipeg Grain Exchange.[23] At the annual meetings of the Territorial and Manitoba Grain Growers' associations early in 1905, he urged the establishment of a producers' grain marketing concern. A joint committee formed by the associations to study the proposal reported to the annual meetings in 1906 in support of the principle of a farmers' grain

[22]See United Grain Growers Limited, *The Grain Growers Record, 1906 to 1943* (Winnipeg, 1944), p. 6; also W. A. Mackintosh, *Agricultural Cooperation in Western Canada* (Kingston, 1924), *passim*; and Patton, *Grain Growers' Co-öperation in Western Canada*, pp. 43 ff.

[23]Partridge spoke of the Winnipeg Grain Exchange as "the house with the closed shutters," and described it as "a combine" with "a gambling hell thrown in." "The wheat business," he said, "is thus practically in the hands of three milling companies and five exporting firms." As cited in Mackintosh, *Agricultural Co-operation in Western Canada*, p. 19.

company to be organized independently of the associations. By that time, much of the preliminary work of organization had been carried on by the Sintaluta group. The purposes of the agency were outlined in a resolution which stated, in part: "That the proposed company, while applying for more extended powers under its charter, shall have for its immediate object the carrying on of a grain commission business or a combined grain commission and track buying business with headquarters at Winnipeg and a seat on the Grain Exchange."[24]

The Grain Growers' Grain Company was incorporated by Manitoba statute in 1906 and began business as a commission firm in the Winnipeg market in September of that year. Its expulsion from the Winnipeg Grain Exchange in November gave tremendous impetus to the western farm movement in terms of support both for its educational or protective associations and for its new commercial organizations, the Grain Growers' Grain Company itself.[25] The opportunities thus created for vigorous protest against the actions of the Exchange served also as occasions for the discussion and expression of protest on various related aspects of the grain marketing problem. The Manitoba Legislature naturally found itself at the centre of the storm since both the Exchange and the farmers' company had been created and continued their respective existence by virtue of charters issued by the Manitoba government.

The Grain Growers' Grain Company first turned directly to the Manitoba government late in December, 1906, after its application for readmission to the Exchange had been rejected. In presenting its application, the Company had insisted that the commission rule which had ostensibly constituted the reason for its expulsion was purely arbitrary. It had nevertheless expressed a willingness to advise its members that patronage dividends were not permissible according to the terms of the charter held under the Manitoba Joint Stock Companies Act, and to consider the matter further at the shareholders' meeting in January, 1907. When the Exchange refused to readmit the Company on this basis, the Company requested the government to insist that the Exchange revise its rules in certain important respects or, if the Exchange refused that, to amend its charter. These requests were accompanied by the specific

---

[24]United Grain Growers Limited, *The Grain Growers Record, 1906 to 1943*, p. 6.

[25]"In conversation with the writer," says Professor Mackintosh, "Mr. Crerar, the President of the Company, and Mr. Kennedy, the Vice-President, expressed the view that the company might never have survived had it not been for the publicity given by expulsion from the Exchange. In attempting to destroy a small farmers' concern, certain grain interests created a formidable rival." Mackintosh, *Agricultural Cooperation in Western Canada*, p. 27 n.

threat that unless they were granted the Company would abandon its trading activities and engage in political agitation to the embarrassment of the government.[26] The Premier, enclosing a copy of the letter from the Company, advised the Exchange to reconsider its actions, suggesting that the Legislature might not have "intended to give the charter and powers [to the Exchange] for the purposes alleged."[27] The Exchange stood upon its legal position and declined. Three months later, in anticipation of a general election, the acting premier informed the Exchange that its actions represented an "arbitrary and unjustifiable exercise of the powers conferred through [its] charter," and threatened legislative action at a special session if necessary. Meanwhile the annual meeting of the Company had ruled against the distribution of patronage dividends, and the Exchange, being informed of this action in a new application for re-admission, restored the Company to the privileges of membership.

The reinstatement of the Company on the terms laid down by the Exchange did not convince organized western farmers that the Exchange, as then constituted, occupied a position in the grain trade acceptable to the producers of grain. On the suggestion of the Manitoba Grain Growers' Association, the provincial Premier called a conference of interested groups in June, 1907, and agreed that his government would implement in law the resolutions of the conference. The meeting was attended by representatives of the Exchange, the Grain Growers' Association, the banks, the railways, and the rural municipalities. The representatives of the Exchange withdrew from the conference in protest against proposals that the government should interfere with its management and the conference then proceeded to enumerate various measures for its supervision. During the session of 1908 the provincial Legislature forced through a bill to amend the charter of the Exchange under threat of governmental resignation. As a result of this action the Exchange dissolved in February, 1908, and in September of the same year reorganized as an unincorporated voluntary association, on which basis it has operated ever since.

The first adequate measures for dealing with the problem of local elevators in western Canada rested in part on provincial assistance. The governments of all the prairie provinces displayed a high degree of responsiveness to agrarian pressure for action. The voice of the farmer was easily recognizable as the voice of the local elector, a person whose presence was obviously far less remote from Winnipeg or Regina or Edmonton than it was from Parliament Hill in Ottawa.

[26]Ibid., p. 25.
[27]Ibid.

The attack of the grain growers on the Exchange had been coupled with the request that the provincial features of the Partridge Plan be implemented.[28] Early in 1907 the Manitoba Grain Growers' Association requested the Manitoba government to undertake the ownership and operation of the country elevators in the province. A resolution from the June conference, agreed to after the withdrawal of the representatives of the Grain Exchange, urged the provincial government "to acquire and operate a complete line of storage elevators throughout the province."[29] At the convention of the Association in 1908, under the guidance of Partridge, a scheme was endorsed for the "provincial ownership and operation of a system of line elevators." The first issue of the *Grain Growers' Guide*, the only one edited by Partridge, was chiefly devoted to the elaboration and sponsorship of the plan.[30]

Growers in the other provinces were obviously interested as well. The Saskatchewan Grain Growers' Association had urged federal construction and operation of interior storage elevators before the federal Commission of 1906, but the Commission advised against the proposal. At its annual meeting in 1907 the Saskatchewan Association approved the principle of government ownership of elevators and passed a resolution requesting the federal government to take over the terminals and to build interior storage elevators. In Alberta the Alberta Farmers' Association urged the construction of a federal elevator at Calgary. The Interprovincial Council of Grain Growers' and Farmers' Associations, which was formed in 1907, urged the executives of the three provincial associations to call upon their respective governments for a declaration of policy.

On the instigation of Premier Roblin of Manitoba, the three premiers met twice in 1908 to consider a course of action. They intimated that provincial action might be taken, but eventually reported to the Council in writing in 1909 that the plan, besides being extremely costly to implement,[31] would be *ultra vires* the provinces since it obviously called upon the provinces "to create a complete and absolute monopoly" of grain

---

[28]The demand which took shape in the period from 1905 to 1910 for the socialization of the elevator system embodied the joint proposals that the Dominion government should take over the terminal elevators and the provincial governments the locals. The scheme came to be known as the "Partridge Plan" after its leading advocate, E. A. Partridge.

[29]Cited in Mackintosh, *Agricultural Cooperation in Western Canada*, p. 35.

[30]The first issue appeared in June, 1908. Partridge resigned after it was published, partly, at least, because his colleagues did not share his views that farmers and labourers should unite for action. See Patton, *Grain Growers' Coöperation in Western Canada*, 70 ff.

[31]They estimated the cost of acquiring the 1,334 licensed elevators in the West at from seven to ten million dollars.

storage facilities and would involve control of the grading and weighing of grain and of the railway companies.³² The communication expressed the willingness of the provincial governments to undertake the programme if the necessary constitutional amendments to the British North America Act could be secured. The premiers of Alberta and Manitoba introduced resolutions into their legislatures to the effect that the Governor General in Council be asked to assume ownership of elevators or, failing that, to provide the necessary constitutional authority to the provincial governments to permit them to do so. The Interprovincial Council protested that its request did not imply legal monopoly of the elevator system. Finally, late in 1909, Premier Scott of Saskatchewan advised premiers Roblin and Rutherford that each province should deal with the situation in its own way. The provincial governments were left to settle with their respective rural electorates as best they might.

The Manitoba government was the only one of the three to establish a governmental system of elevators. Its susceptibility to the importunities of the provincial Grain Growers' Association may have been closely related to political realities. Premier Roblin's Conservative government was anticipating a general election in 1910 and could in no way overlook the results of a by-election in the constituency of Birtle in which a prominent member of the Association had been returned. Whatever the political forces at work, the Minister of Education announced at the annual convention of the Association in December, 1909, that interprovincial negotiations on the elevator question had broken down and that the Manitoba government had accepted "the principle laid down by the Grain Growers' Association of establishing a line of internal grain elevators as a public utility, owned by the public and operated by the public."³³

The government requested assistance from the Grain Growers' Association in drafting the legislation for the new venture but in the negotiations on the elevator question as well as in the final measure which was enacted there were important departures from the Association's recommendations. The Grain Growers advocated that the system be controlled by a commission which should be independent of governmental influence, its members to be nominated by the board of directors of the Grain Growers' Association and removable only by a two-thirds vote of the Legislature or by a decision of the Court of Appeal. The government accepted nominations from the Association but appointed the three-man commission subject to the pleasure of the Lieutenant-Governor in

³²Mackintosh, *Agricultural Cooperation in Western Canada*, p. 36.
³³*Ibid.*, p. 37.

Council. The Grain Growers urged the commission to purchase existing elevators, wherever possible, at an evaluated price and, only where that was impossible, to construct new ones. The elevator bill placed the acquisition and construction of elevators under the control of the Minister of Public Works and established arbitration as the instrument for determining purchase prices where negotiations had failed. The Grain Growers were dissatisfied with the legislation to the extent that they refused at first to have any part in it by way of nominating personnel. Eventually they made nominations but the government, against the advice of the directorate of the Association, persuaded D. W. McCuaig, President of the Association, to be chairman of the commission.

The Manitoba elevator venture was an unqualified fiasco.[34] The government made a capital investment of over one million dollars to acquire 174 country elevators with a capacity of 4,300,000 bushels. Ten of the elevators were newly constructed and the remainder were purchased. The system operated under government control for two crop years, those of 1910 and 1911, or from September 1, 1910, to August 31, 1912. On the first crop year alone, on a turnover of five million bushels of grain, there was an operating loss of upwards of $11,000 and an over-all deficit estimated by the government at $84,000 and by the opposition at $110,000. In May, 1912, the chairman of the commission announced that the government would terminate its elevator activities on August 31. On July 20 the government concluded an agreement with the Grain Growers' Grain Company whereby the latter leased the government elevators at an annual rental of 6 per cent on a capital investment placed at $1,160,000, the government to assume taxes and maintenance.[35] Judging that many of the elevators were poorly located or otherwise unsuitable for operation, the farmers' company operated only 135 of the elevators and left 39 of them idle. After renting for several years the Company began purchasing the elevators from the government a few at a time. In 1927 those remaining in the hands of the government were sold to other grain companies, and the Manitoba experiment in elevator ownership came to an end.[36]

[34]*Ibid.*, pp. 39 ff.; Patton, *Grain Growers' Coöperation in Western Canada*, pp. 88 ff.; United Grain Growers Limited, *The Grain Growers Record, 1906 to 1943*, pp. 13–16.

[35]After the second year of operation by the Grain Growers' Grain Company, American grain interests offered the Manitoba government a much higher rental than the Company was paying. The government refused on the ground that it was preferable to have the elevators operated by the farmers' company. United Grain Growers Limited, *The Grain Growers Record, 1906 to 1943*, p. 16.

[36]*Ibid.*

It is easy to understand the failure of the governmental elevator experiment on a purely operational basis. Sifting fact from political fiction as far as may be done, it is clear that many of the elevators were bought at excessive prices under the arbitration process, that many of them were poorly located and poorly patronized. There is apparently considerable truth in the complaint that, with the Minister of Public Works in charge of procurements, the commission was loaded up with the culls from the elevator system already in existence. As many as three or four elevators were purchased at individual shipping points. The elevators were operated exclusively on a warehousing basis, and their sole source of revenue consisted of an initial handling charge of 1¾ cents per bushel along with trifling amounts for secondary storage. The reason for the failure of the project is not far to seek.

The venture from beginning to end provided the occasion for bitter political recrimination. It was projected in anticipation of a provincial election and was instituted by a Conservative government in fulfilment of its pre-election pledge. Members of the elevator commission stood as the appointed representatives of the government. During 1911 the reciprocity issue was fought out in the national political field, and the Manitoba Grain Growers' Association, many of whose members were Liberals and most if not all of them strong advocates of freer trade, took part unofficially in the Liberal fight for reciprocity.[37] The Roblin government, whether active in the national election or not, represented the party which served to defeat the farmers in their fight for tariff adjustment.

In the provincial Legislature the elevator question was reduced to the level of partisan vituperation. "As the House is aware," said the Premier, "the investment in the purchase of the Elevators was entirely due to the demand made by a certain organization in Manitoba [the Grain Growers' Association] and the action of the Legislature was taken in the belief that the organization in question was prepared to lend full support to the scheme. Unfortunately that has not been done."[38] Liberal spokesmen, however, denied that the farmers were responsible for the purchase of the elevators. They argued that the farmers had always regarded the elevators with suspicion. The Premier replied that the delegation from the Association had insisted on government ownership of elevators and had assured him that it spoke for the farmers of Manitoba. "I have since learned that I was mistaken," continued the Premier, "and here I am willing to admit that I was wrong. I took the voice of the demagogue as

[37]J. Castell Hopkins, ed., *The Canadian Annual Review of Public Affairs*, 1911 (Toronto, 1912), p. 540.
[38]*Ibid.*, 1912, pp. 509–10.

the voice of the public and I consequently made a mistake. The farmers didn't want Government Elevators in this Province. Experience has shown that to be a fact for the reason that they do not patronize them."[39]

The Manitoba venture served to discredit proposals for the operation of country elevator systems as a public utility. It was not due to any such demonstration, however, that the Saskatchewan and Alberta co-operative elevator systems were planned and organized on a basis differing markedly from the governmental system in Manitoba. The origins of the Saskatchewan and Manitoba systems were coincidental in time and their differences were due to basic differences in draftsmanship. The contrasts between the two, while not so great as ordinarily supposed, were, nevertheless, of sufficient importance to suggest considerable differences in political philosophy on the part of the governments which were instrumental in the creation of the systems. We do not refer to the fact that the government which instituted a state elevator system in Manitoba was Conservative while those which instituted co-operative systems in Saskatchewan and Alberta were Liberal. Whatever trace of consistency of meaning these political labels may have in the federal field there is no carryover into the provincial field. On certain points of policy, of course, there may be a high degree of uniformity as between the federal and provincial parties of the same name; on other points none at all.

In the circumstances now under review the proposals put before the Liberal government of Saskatchewan were essentially the same as those made to the Manitoba government. Each government was asked, in effect, to implement the provincial portion of the Partridge Plan which involved socialization of the elevator business, the country elevators by the provinces and the terminals by the Dominion. In terms of political philosophy, neither Liberal nor Conservative provincial parties could expect to find a sympathetic or even tolerant regard for state ownership within the framework of principles to which at least lip service was paid by their national counterparts. Until well beyond the First World War the one well-established collectivist economic principle in the national field in Canada was that which sanctioned the use of public funds and public credit for developmental purposes. Coupled with this principle was the proviso that public ownership and management should be avoided at almost any cost.

Under these circumstances it is apparent that the Manitoba government, which had so consistently served to extend the Dominion developmental philosophy into the provincial sphere in the matter of railroads,

[39]*Ibid.*, p. 510.

was particularly out of character in its eventual reaction to pressures concerning elevators. This is so, not because substantial commitments of public funds were involved but rather because such commitments were coupled with public ownership and control. In contrast to such an aberration on the part of the Manitoba government the reaction of the Saskatchewan government was entirely within the national political tradition. The Saskatchewan co-operative elevator system from this viewpoint represented an ingenious adaptation of the national developmental principle which fully sanctioned the use of public funds and credit in almost any form not requiring ownership by the state.

The Saskatchewan Co-operative Elevator Company was incorporated in March, 1911, by a provincial statute[40] which implemented with but few modifications the recommendations of the Saskatchewan Elevator Commission. Late in 1909 the Saskatchewan Grain Growers' Association presented a petition to the provincial government renewing the common charges of the day that the country elevator business was in the hands of companies which bought and sold or processed and sold grain and which, through their monopoly position, could control both domestic and export prices. The petition sought government-owned storage facilities at country points, arguing in support of the request the opinion that "the only feasible plan for the improvement of the condition of affairs reported is that which has been demanded by the organized farmers of our three Provinces of Manitoba, Saskatchewan and Alberta through their representative associations, namely, that the storage facilities in each province be owned by the Provincial Government and operated under an independent commission as a public utility."[41]

The Secretary of the Saskatchewan Grain Growers' Association appeared before the Select Standing Committee on Agriculture and Municipal Law, to which the petition had been referred, to urge the adoption of the proposals and the institution of a thorough preliminary inquiry. The inquiry was authorized by the Legislature on December 14, 1909, two days before the Manitoba Grain Growers' Association was informed by a provincial minister that the Manitoba government had accepted the principle of government ownership of elevators. On February 28, 1910, the Saskatchewan government appointed a royal commission (the Elevator Commission referred to in chapter VII), to investigate the elevator situation and the proposals put forward for dealing with it. Its members were Robert Magill, Professor of Political

[40]An Act to Incorporate the Saskatchewan Co-operative Elevator Company, Limited, *Statutes of Saskatchewan*, 1 Geo. V, c. 39 (1910–11).
[41]*Report of the Elevator Commission of the Province of Saskatchewan*, p. 9.

Wilbor J Bennett New Deal

FOWKE VC National Policy. The Wheat Eco.

3 Acheson TW Nat. Pol. • The Industrialization

2. HD 9049 W5 C285

Canada -- Economic Conditions

Economy at Dalhousie University, chairman; George Langley, farmer and Member of the Saskatchewan Legislative Assembly; and F. W. Green, Secretary of the Saskatchewan Grain Growers' Association. The Commission's report, dated October 31, 1910, was tabled in the Legislature in January, 1911.

In the course of the inquiry the Commission met with the executive of the Saskatchewan Grain Growers' Association and asked it to prepare a draft bill embodying the Association's proposals. To this request the executive agreed, stating that it would rely for a model either on the bill proposed by the Manitoba Grain Growers' Association or on that of the Manitoba government. At a second meeting, however, the executive confessed that neither of the Manitoba bills was satisfactory and submitted a memorandum in lieu of draft legislation. "The scheme [of the Saskatchewan Grain Growers' Association]," reported the Commission, "is at all events comprehensive, and, considering all its features, it is not surprising that they did not draft a bill to be submitted to the provincial legislature."[42] The proposals went far beyond governmental ownership of country elevators, but in so far as these were concerned the recommendation was that public funds should be used to purchase all existing country elevators in the province except for those owned by the milling companies and therefore unobtainable. Since the latter were in operation at only approximately one-fifth of the shipping points they were not regarded as a serious obstacle, since, with "a virtual monopoly," the proposed organization could adequately control the remaining private houses.

The Royal Commission rejected the Grain Growers' scheme. "Our objections to it are not," it said, "founded upon any opposition to the principle of provincially owned storage. Even though that principle were accepted, this particular scheme of provincial ownership is objectionable."[43] If the government were to attempt to buy sufficient of the existing elevators to secure adequate control, to "buy a monopoly," the terms of purchase and the total cost would involve a prohibitive overhead burden for a new enterprise. If, on the other hand, the government did not seek monopoly but purchased elevators with the idea of competing in the storage field with powerful private interests, it would operate under

[42]Ibid., p. 23.
[43]Ibid., p. 35. The Commission also rejected the Manitoba scheme and, months before the first indications of the financial results of that venture were reported, stated its unanimous opinion that, as far as Saskatchewan was concerned, "a scheme similar to the Manitoba scheme would not be satisfactory to the farmers generally on the one hand and on the other would probably end in financial disaster." Ibid., p. 95.

serious disadvantages. It would not be able to provide for street grain which must be sold locally. It would have no guarantee or even likelihood of sufficient patronage. Its competitors, financially strong and with revenues from the purchase and sale as well as from the storage of grain, would be able to cut handling charges and thus secure the bulk of the business.

While the commissioners disavowed any intention of arguing the merits or demerits of public ownership in general, the worst of the defects in the Grain Growers' proposals were, in the views of the Commission, the commonly recognized handicaps of government enterprise. A system of governmental elevators would, it argued, be in inevitable danger of political influence in regard to appointments and contracts. Were a government unsympathetic to public ownership to come to power it would have "a splendid opportunity" to expose the public elevator system to ridicule and defeat. Mainly, however, the dangers to such a system lay in public attitudes. "The Government would be at a disadvantage," the Commission stated, "arising from the fact universally admitted, that there is a general disposition to exact the utmost possible from the public treasury while not giving the utmost return. This is perhaps the greatest obstacle to the development of public ownership and so long as such disposition is general so long will governments find it difficult to compete in matters commercial or industrial with private corporations."[44] Public ownership of elevators, in the view of the Commission, was merely a worse than average example of a bad variety of business organization. "Advocates of public ownership of public utilities," it concluded, "may well hesitate to rest their case on provincial versus private initial elevators."[45]

Instead of public ownership as the solution to the elevator problem the Commission outlined a plan based on the principle of complete control by the grower. Within that basic tenet the subsidiary but still extremely important consideration was to provide for centralized control combined with the maximum amount of local autonomy and interest in the welfare of the organization. The observations of the Commission had indicated the weakness of independent farmer elevators that had absolute local autonomy and responsibility but did not have integration or centralized control. The Manitoba elevator system, not yet an operational unit, was, in the opinion of the Commission, doomed to failure because it provided centralized ownership and control by the government but nothing in the way of local autonomy or individual responsi-

[44]*Ibid.*, p. 96.
[45]*Ibid.*

bility on the part of the growers. The Saskatchewan solution must avoid the weaknesses of both.

The plan put forward by the Commission and made effective by the Saskatchewan Legislature provided for a "line" of country elevators owned and managed by farmers—with provision for terminals which might be added later—each elevator to be constructed only when local farmers, with a specified minimum of crop acreage, had subscribed to sufficient stock in the company to build the local unit. As for control, it was to be for the system as a whole, the shareholders in each local electing a delegate to the annual meeting on the basis of one-shareholder-one-vote without proxy. The annual meeting was to elect the directorate with each delegate having but one vote. In terms of finance, the fifty-dollar shares were to be available for purchase by agriculturists only, with a maximum holding of twenty shares[46] per shareholder and with restricted privileges of transfer.

Whatever the contrasts between public ownership of elevators as in the Manitoba experiment and farmer ownership as in the Saskatchewan co-operative venture, the differences could scarcely be discovered by looking for the respective sources of funds for the two systems.[47] In the Manitoba system the provincial treasury was, of course, responsible for 100 per cent of the financial requirements. In the Saskatchewan co-operative elevator system, in accordance with the recommendations of the Commission, the provincial government advanced 85 per cent of the fixed capital requirements for country elevators. Shares in the Saskatchewan system were sold with 15 per cent of their par value payable on subscription. The government advanced to the company the remaining 85 per cent of the cost of the local elevators, the loans repayable in twenty years, and as security took a first mortgage on the physical assets of the company with a first claim on the unpaid balance of the capital stock. Additional financial assistance of great importance which was

[46]The Commission recommended a maximum holding per shareholder of ten shares only. *Ibid.*, p. 97.

[47]Contrasting the Manitoba "experiment" in public ownership of elevators with the co-operative plan proposed in its report, the Commission said: "Both plans aim at removing initial storage from the ownership of companies interested in the buying of grain. The one plan aims at ownership by the State and management by the government and the other aims at ownership and management by the growers of the grain. Both plans recognize the strength of the feeling of injustice in the minds of many farmers, both seek to create conditions for the marketing of grain which will give farmers confidence and satisfaction, and *both involve financial aid by the state*. The chief difference between the two plans is that in the one the issue is in the hands of the government while in the other it is in the hands of the farmers themselves, and to this Commission . . . it appears that this difference is in favour of the co-operative plan." *Ibid.*, p. 98. Italics added.

given by the government to the company was its guarantee of the company's line of credit with the bank.[48] In the construction of terminal facilities which the company undertook in the years after 1915, the first house, with a capacity of two and one-half million bushels, was built over the years 1915–18 by the company with its own resources. For an extension of two million bushels' capacity which the company required immediately thereafter, the government advanced half of the cost on a loan basis.[49]

The Alberta government was by no means free from pressure for the socialization of elevators, but the agitation was not as intense in that province as in Manitoba or Saskatchewan. With no decision arrived at in Alberta by 1911, the difficulties of the Manitoba venture were already apparent and the Saskatchewan plan of governmentally underwritten co-operative elevators was in the process of organization. The Elevator Commission of the United Farmers of Alberta proposed a scheme similar to that of Saskatchewan with the suggestion that the Alberta government finance the Grain Growers' Grain Company to enable it to provide the necessary local elevators in the province. The Alberta government declined to extend financial aid to the Grain Growers' Grain Company on the ground that it was an extra-provincial company. In 1913 the Alberta Farmers' Co-operative Elevator Company was established under legislative provisions[50] similar to those which had created the Saskatchewan Co-operative Elevator Company.

Differences between the Alberta and the Saskatchewan co-operative elevator companies which were initially of little importance were increasingly significant in terms of development. The Saskatchewan company concentrated its attention exclusively upon the grain handling business while the Alberta company organized and operated livestock handling and co-operative supply departments. The Alberta legislation required the provincial government to advance 85 per cent of the cost of constructing or acquiring local elevators but did not permit the government to underwrite the company's line of credit with the banks. The Alberta elevator company therefore turned to the Grain Growers' Grain Company for financial support and relied on it for the conduct of its

[48]Permitted by the act of incorporation, s. 27–27a. The government also advanced an organization grant of $7,000. Patton, *Grain Growers' Coöperation in Western Canada*, p. 107.

[49]The new credit was permitted by amendment to the act of incorporation in 1919–20. The amendment was proposed by Hon. George Langley, Minister of Municipal Affairs, a director of the company and its first vice-president. Hon. Charles Dunning, formerly secretary-treasurer and general manager of the company, was Provincial Treasurer in 1920. *Ibid.*, p. 180.

[50]*Statutes of Alberta*, 4 Geo. V, c. 13 (1913).

necessary terminal operations. The Saskatchewan company utilized the Grain Growers' Grain Company for the disposal of its grain in the Winnipeg market for the first year of operation, but thereafter established its own commission department in Winnipeg and later constructed terminal elevators as indicated above.

It is not surprising to find that proposals looking toward some form of union among the three farmers' commercial companies were advanced within a few years' time. By 1915 the Grain Growers' Grain Company voted in annual meeting in favour of federation and the other farmers' companies made tentative proposals in favour, with the prior condition that further study be given the matter. Eventually the Saskatchewan Co-operative Elevator Company decided that continued independence was preferable to union on any attainable basis. The Grain Growers' Grain Company and the Alberta Farmers' Co-operative Elevator Company amalgamated in 1917 to form the United Grain Growers Limited.[51] The latter company continues as one of the effective and powerful farmer-owned companies in western Canada to the present day. The Saskatchewan Co-operative Elevator Company was absorbed into the Saskatchewan Wheat Pool by purchase of facilities in 1926.

The farmers' elevator companies were tremendously successful by almost any test that one might apply. They were profitable, independent, and versatile and demonstrated a remarkable capacity for rapid expansion in the functional areas in which they could best serve the growers' interests. By 1920 the two companies were thoroughly established and experienced in the local and terminal elevator fields and in the commission and export businesses. They operated approximately 650 local elevators, the Saskatchewan Co-operative Elevator Company's line comprising 300 and that of the United Grain Growers 350. These elevators provided competition to the line elevator companies' facilities at upwards of one-half of the grain shipping points in the prairie provinces. The farmers' companies controlled nearly one-third of the terminal capacity at the head of the Lakes.

It is clear that by the early inter-war years the farmers' commercial grain companies had revolutionized the competitive position of the western grain grower in the markets in which he disposed of his cereal products. It would require an exhaustive and difficult analysis to indicate the alteration of circumstances in precise detail. In general, however, the farmers' companies had it within their power by the end of the First World War to set the pattern of elevator services and of price relationships which would be most acceptable to their grower owners.

[51]*Statutes of Canada*, 7 Geo. V, c. 79 (1917).

The companies had, for example, made it a point of consistent policy to diminish the vulnerability of street grain by narrowing the spread between street and other prices. They also made a point of providing special binning facilities as generously as possible. Their competition could not be ignored by the line elevator operators in the matter of determining weights, grades, or dockage.

One of the important aspects of the farmers' elevator companies and their operations was their educational effect. Not all the demands of western grain growers could be met by any elevator system, whether private, co-operative, or state-owned. The refusal of private elevator companies to grant a request or to provide a particular type of service left unanswered the question as to whether the elevator companies *could not*, or merely *would not*, oblige the grower. When the management of the farmer-owned elevator companies informed the grower that such and such a request simply could not be met, the grower was more inclined to take their word for it. This is not to say that the wheat grower became convinced that all was right in the marketing world. His protest in the inter-war years was, in fact, at least as vigorous as anything that had gone before. It does signify, however, that the grower shifted the direction and range of his protest. It came to be recognized that there were distinct limits to the improvement which could be effected in the grain grower's competitive position at either the local or the terminal elevator level. Those limits had been approached by 1920.

A further point to notice concerning the farmers' commercial grain companies was that they adapted themselves to the open market system, they accepted its free-enterprise premises and became a respectable and orthodox part of the open market process for the marketing of grain. This point is of particular significance in relation to agrarian protest in the 1920's when, for the first time, the open market system itself came to be suspect. The farmers' elevator companies were accorded their full share of this suspicion and were held to be inadequate to the demands of the new, post-war situation.

One final reference concerns provincial policy and the distinction between state enterprise and co-operative action. The Manitoba elevator scheme stands as an example—and an ominous one—of state enterprise, while the elevator schemes of Saskatchewan and Alberta stand as illustrations of farmers' co-operative activity. It must be mentioned again by way of emphasis, however, that the provincial governments materially assisted the elevator systems in Saskatchewan and Alberta. It is no derogation of the unquestioned managerial skill displayed by the farmer-owned elevator companies to give due weight to the financial aid of the provincial governments in explaining the outstanding achievements of these companies. As Professor Mackintosh said in the middle twenties, "Undoubtedly the aid of the government has been a powerful factor in the success of the Saskatchewan company."[52] The remark would apply to the Alberta company as well. The individual co-operative elevators described earlier in this chapter, and the Grain Growers' Grain Company, offer the only examples of autonomous and independent, co-operative, grass-roots effort directed toward the marketing of western Canadian grains prior to 1920.

There are obvious valid distinctions between socialism and co-operative effort, and between both of these and ordinary private enterprise. Yet much confusion of thought must result from an analysis based on the assumption that these three forms of business organization have nothing in common and must universally and in their entirety be mutually exclusive. Of all the levellers of distinctions in the economic field, credit is, perhaps, the greatest of all. If governments provide the bulk of the capital required by co-operative and private-enterprise organizations it may become difficult to differentiate sharply between the two and to distinguish them from state enterprise. This point ought to be kept in mind in any comparison between the reactions of the Manitoba government and those of the governments of Saskatchewan and Alberta to farmers' agitation.

Professor Mackintosh has put on record another significant comment regarding the co-operative elevator companies and their relation to the provincial governments: "The justification of the action of the Saskatchewan government in thus aiding the co-operative company is to be found in the fact that the political alternative to this project was government ownership of country elevators." An overwhelming desire to avoid the taint of socialism has been one of the persistent realities of Canadian political and economic life down to the present day.[53] Governments at the federal and provincial level, at least, have of necessity exercised their ingenuity to discover ways of regulating the rate and direction of economic activity in their constituencies without hazarding their free-enterprise virtue or good name. The Dominion government has favoured the subsidization of private enterprise, as in railways and iron and steel.

[52]Mackintosh, *Agricultural Cooperation in Western Canada*, p. 76.

[53]Not excluding the political life of Saskatchewan under the C.C.F. government. For a provocative analysis of this segment of Canadian political thought see S. M. Lipset, *Agrarian Socialism: The Co-operative Commonwealth Federation in Saskatchewan* (Berkeley and Los Angeles, 1950), particularly chaps. VI, VII.

The governments of the prairie provinces have to a considerable extent sponsored co-operative activity, as the Saskatchewan government did the elevators, as the "political alternative to government ownership."

The question which can be raised but not answered here, however, is how much financial assistance can the state give to a private enterprise or to a co-operative organization without altering the distinctive character of these respective types of economic agencies? Assuming continued financial competence on the part of the agency in question, the common answer would appear to be that state aid is a matter of indifference. The Canadian Pacific Railway Company has never, since its earliest days, been seriously threatened with insolvency. It has accordingly been characterized as private enterprise without question and without reference to its subsidized status. Similarly the Saskatchewan Co-operative Elevator Company prospered and fully met its fixed charge commitments to the Saskatchewan government. It was never regarded as a form of state or quasi-state activity. The question of distinctive characteristics is, however, sharply raised in circumstances also illustrated in Saskatchewan, where a state-financed co-operative company—the Saskatchewan Co-operative Creameries—did not prosper or even attain solvency. Under these circumstances the designation of the company as a co-operative was not sufficient over the decades to save the government from persistent responsibility and recurrent embarrassment.

CHAPTER NINE

Western Protest and Dominion Policy

LOCAL AND TERMINAL ELEVATORS, along with railways, formed the costly but indispensable physical equipment for the marketing of western grain. The Dominion government had no philosophical aversion to monopolistic control either of elevator or railway facilities, but by 1900 it was possessed of a firmly established pragmatic aversion to monopoly of any sort which threatened to interfere with western development or with national policy. Despite the cancellation of the monopoly clause, the Canadian Pacific Railway Company inevitably retained a strong measure of monopoly within the area contiguous to its lines and shipping points. Toward the end of the century the railway had extended the monopoly privilege to the elevator field by agreeing with any party who would build a standard elevator at a loading point that it would prohibit the loading of cars at that point either over the loading platform or through a flat warehouse. Agrarian protest against this practice, voiced in the House of Commons, prompted governmental investigation and regulation. Legislative controls over grain trading agencies were immeasurably broadened after 1900 by new legislation and by repeated amendment to the old.

Legislative control in the grain trade was by no means new. Governmental inspection and grading of grain as one of the groups of "staple articles of Canadian produce" was well established in Canada prior to Confederation. It was translated into federal law in a general inspection act of 1873,[1] and, through repeated revision, this act was kept in conformity with changing circumstances. The revisions of 1885, 1889, 1891, and 1899[2] effected the necessary legislative transition to harmonize with the shift from an eastern grain trade in winter wheat to a national trade in spring wheat of western origin. Various grades of Manitoba "hard" wheat headed the list of descriptions in the revision of 1885. The amendments of 1889 and 1891 provided for the establishment of a western as well as an eastern standards board, and for the establishment of "commercial" grades for western grain, these grades to be applicable

---

[1]*Statutes of Canada*, 36 Vic., c. 49 (1873).
[2]*Ibid.*, 48–49 Vic., c. 66 (1885); 52 Vic., c. 16 (1889); 54–55 Vic., c. 48 (1891); and 62–63 Vic., c. 25 (1899).

in cases where crop conditions prevented the inclusion of a considerable proportion of the crop in the statutory grades. The General Inspection Act of 1899 created the inspection district of Manitoba, defined to include Manitoba, the Northwest Territories, and Ontario as far east as the head of the Lakes.

Changes in the legislative provision for inspection and grading, instituted in response to the petitions of boards of trade and grain trading interests,[3] went a long way to improve the Canadian system of the sale of grain by official grade. They did not prevent the recurrence of farmers' criticism in the matter of inspection and they did not even touch the question of governmental control of grain handling agencies.

In 1898 James Douglas, member for East Assiniboia, introduced a bill into the federal House "to regulate the shipping of grain by railway companies in Manitoba and the North West Territories." The main purpose of the bill was to secure for farmers the legal right to load grain over loading platforms or through flat warehouses. Its premise was that the powerful element of monopoly inherent in the position of the Canadian Pacific Railway Company—even after the cancellation of the monopoly clause—was being extended to the elevator field by agreement between the railway and elevator companies, and must be broken. The bill made no progress in the 1898 session but was reintroduced in 1899. Its original introduction prompted the Canadian Pacific Railway Company to provide cars for platform loading for the crop of 1898. Discussion of the bill before a special committee of the House in 1899 led the government to appoint a royal commission "on the Shipment and Transportation of Grain in Manitoba and the North-West Territories."

The Commission of 1899 was the first of half a dozen federal royal commissions to investigate the grain trade of western Canada within the next forty years,[4] and the first of two appointed before 1920. In 1906 a farmers' delegation appeared before the Agricultural Committee of the House of Commons to protest the inadequacy and injustice of the inspection and grading of western grain as carried on under existing legislation. On the basis of requests thus presented, the federal government appointed a second royal commission to investigate the farmers' com-

---

[3]See V. C. Fowke, *Canadian Agricultural Policy* (Toronto, 1946), p. 243.

[4]Commissions were appointed in 1899, 1906, 1923, 1931, and 1936. An additional commission was appointed by the federal government in 1921 but its investigations were halted by an injunction sought and secured by the United Grain Growers Limited and upheld by the Supreme Court of Canada. For the citation of the reports of these commissions seriatim see V. C. Fowke, "Royal Commissions and Canadian Agricultural Policy," *Canadian Journal of Economics and Political Science*, vol. XIV, no. 2, May, 1948, p. 165 n.

plaints and to recommend changes in the Grain Inspection Act and the Manitoba Grain Act. Since royal commissions were used repeatedly by Dominion and provincial governments after 1900 in the development of policy concerning the wheat economy, and since such investigating bodies may have a variety of uses, it is well to note certain features of these two commissions. This is of particular importance because of contrasts in regard to the type and purpose of the royal commissions appointed by the federal government before and after 1920.

The chief point of interest concerning the two federal commissions of 1899 and 1906 is that their personnel was made up almost exclusively of western grain growers.[5] The central purpose of the commissions was, it may be presumed, to hear and evaluate the complex of grain growers' complaints concerning grain handling facilities. Yet no attempt was made by the government to observe, in the selection of personnel, any of the usual rules concerning a proper representation of the various parties vitally interested in the dispute, or of the various geographic areas of the national economy. The Commission of 1899 consisted of three Manitoba farmers under the chairmanship of Judge Senkler of St. Catharines, Ontario. When Judge Senkler retired on account of ill health he was succeeded by A. E. Richards (afterwards Mr. Justice Richards) of Winnipeg. It is clear from the records that C. C. Castle, one of the farmer members, dominated the group. The Commission of 1906 was made up of three western farmers including the chairman, John Millar of Indian Head.

The facts noted above suggest that the early federal commissions on the grain trade were not regarded as instruments for the impartial evaluation of agrarian protest against the elevator and milling companies and the railways. They also suggest that the governments which appointed them were not seeking to weight the scales against the protesting western wheat growers, and that there was no attempt to select royal commissioners who could be counted on to reduce the farmers' protests to the level of ridiculous tirade by the refinements of economic sophistry. The Dominion government was, apparently, prepared to take seriously the wheat growers' protests and was not, in fact, willing to run the risk of having these protests effectively countered by the presence of representatives of any of the defendant interests. This characteristic of royal commission membership before 1920 was not typical of the commissions appointed by the Dominion government to investigate the problems of the West. The membership of such royal commissions after 1920 was of a different character and this difference provides one of

[5]For a comparison with the inter-war commissions, see chap. x, pp. 188 ff.

the sharpest of the contrasts between the policy-making devices employed in the two periods respectively.

The partisan nature of the membership of the pre-1920 royal commissions on the grain trade makes it clear that they were not appointed primarily as impartial fact-finding bodies. The facts and the fancies which were likely to be encountered were well known and the commissions were used as a device for putting them on the record. In doing this, they were to serve two purposes: they were to perform the well-recognized safety-valve function so common to royal commissions; and they were to educate the public and the Dominion Legislature on the necessity for a certain type of legislative enactment. The first purpose is self-explanatory. The explosive pressures which the commissions were designed to release harmlessly are readily apparent in the record of the history of western development. The second purpose requires a brief comment. Its implication is that the royal commissions of 1899 and 1906 were not so much expected to formulate policy or even to advise on the formulation of policy as they were to establish the need and rationale for policy already awaiting application. They recommended a single solution to the problems created by the existence of monopolistic elements in the grain trade over the turn of the century. Their recommendations called only for licensing and supervision by a permanent governmental agency.

The royal commission has too well earned the reputation in Canada of being an excellent alternative to effective governmental action, an instrument to justify procrastination. While the governmental use of later royal commissions on the Canadian grain trade may have contributed to that reputation, this was by no means true of the commissions of 1899 and 1906. The commissioners applied themselves diligently to the recording of information relevant to their terms of reference, they reported their recommendations in some detail, and the government promptly implemented these recommendations by legislation—each time, it may be said, on the eve of a general election.

Reporting on its investigations of the charge that local elevators had secured a monopoly position by the aid of the railways, the Commission of 1899 indicated that the refusal of the railways to accept grain from flat warehouses or to furnish cars directly to farmers had forced the growers either to ship their grain through the elevators or to sell to them. The resultant "lack of competition," the Commission stated, had given to elevator owners the power to depress prices. "It would naturally be to their interest," the Commission added, "to so depress prices; and when buying to dock as much as possible."[6] The only adequate remedy for the specific failure of competition at local shipping points, it appeared to the Commission, lay in the fullest possible freedom of shipping. It therefore advised that the railways be required by law: (1) to permit the construction of flat warehouses at loading points; (2) to construct loading platforms themselves on the basis of farmers' petitions; and (3) to provide farmers with cars "as a legal right" and not as a matter of privilege. For the proper regulation of the trade in general it recommended the adoption by the Dominion government of the system already well established in Minnesota, where a state commission, the Railroad and Warehouse Commission, was responsible for the administration of a general supervisory act relating to the shipment of grain.

The Manitoba Grain Act[7] was passed by the Dominion government in 1900 to implement the recommendations of the Commission. Apart from the general question of grain trade regulation, the major emphasis was on the provision of alternative opportunities for the disposal of grain at local points in order that some effective degree of competition might be restored. This emphasis, so pronounced in western protest, had been translated directly into the recommendations of the Commission and was the basis of the central legislative requirements of the act of 1900. It provided for the licensing, bonding, and supervision of all grain dealers, and established the office of warehouse commissioner for administration.[8] It required the railways to supply cars without discrimination for loading through flat warehouses or elevators, or over loading platforms. Furthermore, the railways were required, on the receipt of a written application of ten local farmers with a certain minimum acreage in crop, to construct a loading platform and to make it available without charge. Standard forms were established for recording the various transactions involved in the transfer, storage, and sale of western grains.

Neither the entire act nor any one of its sections would be of benefit to grain growers if interested parties were to be allowed to circumvent the legislation passively or in open defiance. Construction of loading platforms and permission for the erection of flat warehouses as required by the act were indispensable to the creation of alternative local oppor-

[6]Canada, *Sessional Papers*, 1900, no. 81a, p. 10.

[7]*Statutes of Canada*, 63–64 Vic., c. 39 (1900).

[8]C. C. Castle, one of the farmer members of the Commission of 1899, was appointed first warehouse commissioner.

tunities for the disposal of grain. These provisions, however, could be frustrated in their purpose unless they were accompanied by the assurance that cars would be provided with complete impartiality for loading through the various agencies, so the car distribution clause (s. 44) proved to be of crucial importance in the implementation of the federal policy of grain trade regulation in the early years of the century. Adequate observance of this legislative requirement was not established until the Sintaluta test case of 1902 wherein the agent of the Canadian Pacific Railway Company at Sintaluta was found guilty of a breach of the relevant section of the act. The decision was upheld by the Supreme Court of Canada on appeal and the validity of the law was thus made certain. Remaining ambiguities in the car distribution clause were further reduced by a legislative amendment in 1902 which embodied new wording almost identical with that recommended by the Territorial Grain Growers' Association.[9] The Manitoba Grain Act and the Sintaluta judgment fell far short of a complete removal of the dissatisfaction of grain growers with local shipping conditions but they went a considerable distance in that direction.

Shortly after the turn of the century the attention of western growers was shifted increasingly to the terminal market and to the activities carried on in terminal elevators at the head of the Lakes, to the processes of inspection and grading in Winnipeg, and to the conduct of business in the Winnipeg Grain and Produce Exchange. In response to a request from the Territorial Grain Growers' Association, the Territorial Department of Agriculture had milling and baking tests conducted at the Ontario Agricultural College, Guelph, in 1903 and 1904, to determine the bread-making qualities of the various grades of western grain. The report of the expert showed variations in baking qualities which had little relation to typical price spreads between the grades.

The publication of the results of these investigations intensified the demands for further changes in the system of governmental inspection and grading. The federal Grain Inspection Act of 1904 consolidated existing legislation in this field but dissatisfaction continued unabated. The Grain Growers' Grain Company, organized in 1906, sought to operate within the established system and to break the monopoly elements which appeared to characterize the private trade.[10] What a farmers' commission agency might accomplish in the terminal market was far from clear in 1906. Certain points, however, were evident. Such

[9]See H. S. Patton, *Grain Growers' Coöperation in Western Canada* (Cambridge, Mass., 1928), pp. 34–5.
[10]See chapter VIII, pp. 136–8.

an agency could neither alter the inspection system appreciably nor, without terminal elevators of its own, could it modify the method of operating the terminal elevators. The farmers' delegation that appeared before the Agricultural Committee of the Dominion House in 1906 protested particularly against the inspection and grading system. The royal commission appointed following these representations was in-structed "to take into consideration all or any matters connected with the Grain Inspection Act and the Manitoba Grain Act. . . ."[11] The Commission, reporting in 1908, recommended approximately fifty specific amendments to these laws in order that they might constitute "a more thorough system of supervision and control." The Dominion government promptly implemented the recommendations *in toto.*

The reaction of the Commission of 1906 to the question of ownership of elevators by the Dominion government is of particular interest. The question was raised in a preliminary way at the hearings and was summarily dismissed. The proposal put forward by the Grain Growers' associations of Manitoba and Saskatchewan was that the Dominion government should construct and operate a number of conveniently located interior terminal elevators. It was argued that such facilities would relieve car shortages by shortening the turn-around time and would provide growers with official grade and weight with a minimum of delay. The Commission advised against the expedient as excessively costly and one which would involve shippers in unnecessary expense. Its opinion was that the option of shipping to an interior terminal was one that few shippers would choose since "spot grain was the objective except for local mills."[12]

The associations also urged that the Dominion government take over the terminals and operate them as a public utility. On this point the Commission reported as follows:

[11]From Order in Council appointing the royal commission, July 26, 1906 (*Report of the Royal Commission on the Grain Trade of Canada, 1906*, Canada, *Sessional Papers*, 1908, no. 59, p. 3). That the investigation was meant to be thorough is indicated by the power extended to the Commission "to visit the grain growers, the elevators all over the wheat-growing region, the methods of handling the grain at the various stations, farmers' elevators, as well as companies' elevators, the distribution of cars, methods of the grain dealers in Winnipeg, Toronto and Montreal, and the system of government inspection and collection of fees, selection of grades and the methods of handling the grain at Fort William and Port Arthur, at the lake ports, at Montreal, St. John and Halifax, and also the conditions existing as to the manner of handling the grain upon its arrival in England" (*ibid.*). The commission conducted its investigations in western and eastern Canada, in the United States, and at a dozen centres in the British Isles.

[12]*Ibid.*, p. 14.

To prevent the evils that are made possible by operation of the terminal elevators under the present system, we do not think it wise to advise the Government to go to the length of taking over the terminal elevators or of prohibiting persons engaged in the grain trade being interested in such terminals. We believe it possible to obtain a good service from these elevators under the present ownership by having a more thorough system of supervision and control.[13]

The three farmer commissioners of the 1906 investigating group were clearly out of sympathy with the representatives of the Grain Growers' associations in regard to the question of socializing the terminals. They were, however, in harmony with the views of the Dominion government of that day and those of many years to come. The views which they expressed may also have been acceptable to a large majority of western grain growers at the time, for the agitation for government ownership of grain handling facilities was as yet only in its initial stages.

Although the Canadian Pacific Railway Company did nothing by way of providing local grain buyers with elevator facilities, it did construct terminals at the head of the Lakes. From 1884, when it built the first terminal at Port Arthur, until 1904, the terminal facilities at Fort William–Port Arthur were exclusively operated by the Company. In 1902 the Canadian Northern Railway, just beginning its expansion throughout the West, built large terminals at the lakehead, which it leased to grain companies in 1906. In 1904 two of the grain companies, the Ogilvie Flour Milling Company and the Empire Elevator Company built public terminals.[14] Thus, after 1904, the terminals gradually came into the hands of the grain companies and became part of their integrated systems.

Integration, both regional and functional, has characterized Canadian grain handling agencies from the early days. Companies which served as warehousemen through the ownership of local and terminal elevators ordinarily bought and sold grain and thus acted as merchants also. Farmers came increasingly after the turn of the century to believe that this combination of functions was detrimental to their interests, particularly when it occurred in the terminal field. It was considered that a terminal operator who bought and sold grain on his own account could not be expected to offer disinterested service to patrons whose grain he was called on to handle on a storage or transfer basis. The chief charges against such operators were those of mixing and undercleaning of grain, both of which practices allegedly had the effect of diluting Canadian

[13]*Ibid.*, Appendix E, p. 39.
[14]See *Report of the Royal Grain Inquiry Commission* (Ottawa, 1925), p. 76.

export grain to the minimum of each respective grade and thus of lowering its reputation as well as its price in the overseas markets.

With these circumstances in mind it is not difficult to realize that the complaints of the western grain growers were not silenced by the "more thorough system of supervision and control" of the elevator system which was provided in 1908 by amendment to the Grain Inspection Act and the Manitoba Grain Act in conformity with the recommendations of the federal Royal Commission of 1906. In May, 1909, a delegation from the Interprovincial Council of Grain Growers' and Farmers' Associations urged federal operation of terminal elevators before Sir Richard Cartwright, federal Minister of Trade of Commerce. In this representation the farmers' delegation was supported by the Dominion Millers' Association. Investigations conducted officially by Commissioner Castle and independently by the Manitoba Grain Growers' Association late in 1909 confirmed the long-standing allegations of the farmers that terminal elevator operators mixed and undercleaned the grain which they handled. The views of officials of western farm organizations were summed up in an editorial in the *Grain Growers' Guide* of December 28, 1909: "Just so long as these [terminal] elevators remain in private hands, there will be the temptation to private gain. There is only one possible method by which the system of robbing the farmers' grain at the terminals can be abolished. That method is by federal government ownership."[15]

The agitation continued unabated throughout 1910. In January the Manitoba Grain Growers' Association presented a memorandum to Cartwright embodying the same proposal as before. On this occasion the Association was supported by representatives of eastern millers and by a number of independent grain dealers of Winnipeg. Among the various demands presented repeatedly to Sir Wilfrid Laurier on his western tour of 1910 was the demand for federal operation of terminals and for the prevention of mixing therein. The Prime Minister expressed his willingness to discuss the terminal matter further, and in October, on his return to the capital, he wrote to the Grain Growers' associations saying that the government would hear their representations. The associations decided against further separate appearances in this matter in view of the march on Ottawa planned by the Canadian Council of Agriculture.[16] One of the major petitions presented by the massive

[15]As cited in Patton, *Grain Growers' Coöperation in Western Canada*, p. 135 n.
[16]The Canadian Council of Agriculture was formed in February, 1910. D. W McCuaig, president of the Manitoba Grain Growers' Association, was its first president.

delegation of 500 western grain growers and 300 eastern Grangers in December, 1910, was, therefore, a repetition of the demand for federal operation of terminals.[17] Additional support was given to this memorial by the Dominion Millers' Association, the Toronto Board of Trade, and eastern and western exporters.

The representations of 1910 may be regarded as the climax of the pressure put on the federal government to operate the terminal elevators. The government refused to yield except by way of partial compromise. The well-established policy of attacking abuses by legislative control and improved supervision was continued in the Canada Grain Act which was introduced by the Liberals in 1911 and enacted by the Conservative government in 1912. This act consolidated the Grain Inspection and Manitoba Grain acts, it replaced the office of warehouse commissioner with a board of grain commissioners and provided for the mixing of grain in "hospital elevators." As a concession to the persistent western demands for socialization of the terminals the act provided for the construction or acquisition of terminals at the head of the Lakes by the Dominion government and for the operation of such elevators by the Board of Grain Commissioners.[18] The government built a terminal at Port Arthur in 1913 with a capacity of 3¼ million bushels and, on the recommendation of the Board of Grain Commissioners, constructed interior terminal elevators at Moose Jaw, Calgary, and Saskatoon beginning in 1913–14, and, a decade later, one in Edmonton.[19] These elevators were placed under the management of the Board of Grain Commissioners.

[17]Other demands made by the delegation were for tariff reductions, increased British preference, reciprocity with the United States, completion and operation of the Hudson Bay Railway and its terminals by the government, the establishment of a chilled-meat industry, federal co-operative legislation, etc.

[18]In outlining the intentions of the government in relation to the section of the grain bill which provided for government operation of terminal elevators, Hon. Mr. Foster said: ". . . we do not intend to undertake the financial or experimental responsibility of taking the whole terminal elevator system under government operation for the present, but we wish to give to the people of the West a choice between the terminal elevators that are run by corporations or individuals and those that are run by the government either as owners or lessees." Canada, *House of Commons Debates*, Dec. 28, 1912, as cited in Patton, *Grain Growers' Co-öperation in Western Canada*, p. 145.

[19]For an analysis of the reasoning on which the construction of the interior terminal elevators was urged and accomplished see R. Magill, *Grain Inspection in Canada* (Ottawa, 1914), pp. 54–8. For an analysis of the extent to which these elevators achieved their purposes on the basis of their first decade of operation see *Report of the Royal Grain Inquiry Commission, 1925* (Ottawa, 1925), pp. 42–3.

The mixing of grain in terminal elevators remained a troublesome matter for years after the passage of the Canada Grain Act but it, along with the more general question of terminal operation and control, assumed less and less relative importance after 1912. A number of factors contributed to this result. The elaboration of control and supervision provided for by the Grain Act inspired greater confidence. The government's lakehead terminal provided an alternative channel for western grain. More important still, however, was the fact that the terminal field was no longer controlled exclusively by non-farm interests. In October, 1912, the Grain Growers' Grain Company, which until that time had operated as a terminal commission agency without elevator facilities, leased a terminal at Fort William from the Canadian Pacific Railway Company and immediately assumed operation.[20] In January of the following year it purchased another terminal at the same place. Presently the Saskatchewan and Alberta Co-operative Elevator companies were organized and the former company entered actively into terminal ownership and operation. The truth of the matter is that the farmers lost much of their aversion to a number of the practices of the terminal elevator companies by becoming terminal elevator operators themselves. This is notably true of the long-protested practice of mixing grain. The royal commission which reported on the Canadian grain trade in 1925 pointed out that the farmers' grain marketing companies of the day, the Saskatchewan Co-operative Elevator Company, the United Grain Growers Limited, and the three provincial wheat pools with their joint Central Selling Agency, were all engaged in the mixing business.[21]

By 1914, the Dominion government was well on the way toward the fulfilment of the national policy in terms of western economic development, an achievement made with but slight modification of the elements

[20]United Grain Growers Limited, *The Grain Growers Record, 1906 to 1943* (Winnipeg, 1944), pp. 15–16.

[21]It cannot be inferred from this that there was general approval, among the membership of these companies, of the practice of mixing. "Hon. J. A. Maharg, then president of the Saskatchewan Grain Growers' Association, and a director of the Saskatchewan Co-operative, of which company he is now president, said that the directorate of his company, with perhaps one or two exceptions, were individually opposed to the practice of mixing, but because the practice had become prevalent, the company found that they had to go into it to make money to compete with their competitors, and they were forced into it. His personal view was that mixing was done by the farmer on the farm, it was done in the country elevator, it was done by all who handled wheat, and he believed Canadian wheat to be mixed in the United States; and, therefore, since it could not be stopped, mixing in the private elevators at the head of the lakes should not be stopped." *Report of the Royal Grain Inquiry Commission*, 1925, p. 105. For an extended discussion of the mixing problem see *ibid.*, pp. 75–109.

of that policy as originally evolved. Western lands had contributed substantially to the financing of the first transcontinental railway and had offered the major attraction for settlers to become established in the western provinces. Tariff policy remained essentially unchanged from that of the firm protection which had been made effective a generation earlier. The British preferential schedules, adopted over the turn of the century, modified this system to a certain extent, but they made no breach in the concept of an east–west transportation universe of which the St. Lawrence community formed an integral part. The view that a shortage of private capital or a reluctance on the part of private investors should not be allowed to prevent major development had cost the Dominion heavy expenditures for the first transcontinental railway, but had not involved the taint of socialism. Similarly, for the second and third transcontinental systems, the Dominion government had assumed tremendous commitments for security guarantees and for the construction of the National Transcontinental from Moncton to Winnipeg. The free-enterprise system of railway construction was nevertheless regarded as still intact. Monopoly in the railway field, originally regarded as essential to the national policy, was found to be incompatible with that policy and had given way to governmentally sponsored competition and governmentally imposed regulation.

The difficulty encountered in marketing western grain did not appear to be primarily due to a shortage of developmental capital. Elevator facilities were inadequate before the turn of the century, but they were rapidly multiplied thereafter by private capital until in 1910 a royal commission could report that country elevators existed greatly in excess of requirements throughout the West. Western farmers' demands after 1905 were not for governmental assistance in the building of elevators in order to overcome a shortage of such facilities, but, rather, for outright socialization of the local and terminal elevator system of the West in order to break the private monopoly elements which appeared to dominate its activities. Farmers appointed as royal commissioners were able to tell the Dominion government what it was reassured to hear, that adequate regulation rather than socialization was the remedy needed for the real and imaginary abuses of the existing elevator system. In spite of direct and persistent agricultural representations to the contrary in the years 1909 and 1910, the government maintained this conviction and policy and renewed its pledge to that effect in the Canada Grain Act of 1912 as well as in later revisions of that act.

The outbreak of war in 1914 created the emergency which for the first time in the history of the Dominion placed the national military obligation foremost regardless of possible incompatibility with national economic development. The pace of western expansion had faltered before the war. Economic conditions were uncertain. The outbreak of war reduced immigration to negligible proportions and impeded the entry of capital. The country's energies turned from the promotion of immigration and economic development to the solution of the manifold problems of recruitment and military training.

It became evident, as the crisis of the war became more clearly defined, that the economic development of the West in the tradition of the national policy was not contradictory to the requirements of the new situation but was, on the contrary, of the utmost positive importance to it. The popular slogan came to be, "Food will win the war." Because of the problems of shipping, food in this context meant wheat from non-European countries but, more specifically, from countries in the Northern Hemisphere. While further immigration for the occupation of new lands was out of the question for the time being at least, there were in the Canadian wheat economy millions of acres of good land which were occupied but not as yet under cultivation. A major part of the Canadian contribution to the Allied effort in the First World War consisted of a maximization of wheat exports and the fostering of the agricultural effort which made such exports possible.

The Dominion government entered the war and, for the most part, fought through it to the end without seriously questioning the assumption that economic policies which sufficed for peace would suffice for war.[22] Essentially the belief was in "business as usual" or in business according to free-enterprise principles for the allocation of the factors of production. Price served as a satisfactory guide for the allocation of labour, capital, and land in peacetime. The use of the factors of production might well be guided by the same directive during war, with one modification—patriotism. Price and patriotism, so it was argued, would adequately serve to guide the economy and determine its goals. "Recruit rather than conscript" expressed the economic as well as the military philosophy of the Dominion government as, indeed, of the Allies generally.

Modification of this viewpoint was forced upon the Canadian as upon other Allied governments. For the Dominion government at least it was a modification founded on expediency rather than on principle,

[22]This attitude was in sharp contrast to the approach of the Dominion government to the Second World War; substantial measures of governmental control were drafted before the fall of 1939 in anticipation of the war and other restrictive measures were matured and made effective month by month as the war progressed.

and an alteration to be reversed as quickly as possible. In regard to manpower, conscription became essential, but only for military purposes and for the latter part of the war. In railways, a chaotic situation, with no possibility that the two incomplete transcontinental systems could either be left as they were, or completed and operated on the basis of further governmental guarantees, compelled the eventual nationalization of more than half the Canadian railway milage. This measure occurred without hope of later reversal but also without the acceptance by the government of the principle that nationalization was anything but an undesirable necessity. As to the market structure for goods and services other than those for the military and for a part of the railway system, few modifications were regarded as necessary at any time during the war—with the eventual exception of wheat.

Even in regard to the marketing of Canadian wheat the traditional laissez-faire policy was maintained for upwards of three of the four war years. The only changes demanded of the wheat economy were quantitative ones. It was a question of more and more. Western Canada was a wheat exporter of marked significance to western Europe well before 1914 and the prompt exclusion of Russia from western markets by the closing of the Dardanelles early in the war focused Allied demands on North America. In contrast to the food policies developed during the Second World War, the Allies' demands for food were concentrated on wheat on a general assumption that an effective war effort required above all that bread be plentiful and cheap.[23] The major shift in the world demand for wheat was, however, left for a considerable time to work itself out through ordinary market channels. While governmental purchasing was established in Italy, the grain trade in Britain and France was left strictly in private hands. Cash and futures markets remained open, and private importers, millers, and other agencies carried on the trade as before. With the market system left to its own devices in the major importing countries and in the United States, it would have been surprising to find any early modification in the Canadian grain marketing system, the "open market" system with its focal point in the Winnipeg Grain Exchange. For two years after the outbreak of the war the Allies secured necessary foodstuffs through the ordinary import channels in reasonably ample quantities and without pronounced

[23]The British Royal Commission on Wheat Supplies stated that bread was "the only diet which sufficed in isolation and was therefore, indispensable" (as cited in Mitchell W. Sharp, "Allied Wheat Buying in Relationship to Canadian Marketing Policy, 1914–18," *Canadian Journal of Economics and Political Science*, vol. VI, no. 3, Aug., 1940, p. 372).

advances in price.[24] The Canadian bumper wheat crop of 393.5 million bushels in 1915, grown on an acreage increased by five million acres over that of 1914, contributed materially toward that end.[25]

However, drastic alterations in supply and price during the summer months of 1916 compelled an alteration in food policy, first concerning Allied procurement and, second, by resultant pressures, in the overseas producing countries. Under conditions of unrestricted submarine warfare, the price of wheat in Britain advanced by one-half from June to October, 1916. On October 10 the Liverpool futures market was closed and the British Royal Commission on Wheat Supplies was appointed as an agency for investigation and for control of the procurement, first of wheat and flour, and, shortly after, of all cereals. By the end of November, 1916, the United Kingdom, France, and Italy had signed the Wheat Executive Agreement to purchase all wheat requirements jointly, under the agency of the British Royal Commission on Wheat Supplies. Private trading in wheat and flour virtually disappeared in the three countries. The predominant position of the single purchasing agency forced the purchasers of other countries into a position of almost total dependence. Greece joined the buying pool in April, 1917, Portugal and Belgium at a later date, and eventually Norway, Sweden, Holland, Iceland, and Switzerland came to rely heavily on the Wheat Executive of the Allies for supplies of wheat and flour.

Centralized purchasing on the part of the Allies was projected into overseas export markets by the establishment of agencies to secure wheat for the Commission. In the United States the Wheat Export Company, a buying and exporting agency, was incorporated to secure wheat for the Wheat Executive, and the Wheat Export Company of Canada was incorporated under Dominion charter for the same purpose. These companies replaced the regular exporters in the respective countries. For a number of months, however, the regular grain markets remained open in Canada and the United States and the Wheat Export companies purchased in cash and futures markets alike as ordinary buyers. It is clear that while the markets remained formally unchanged, their structure

---

[24]A single exception to the ordinary open market procurement of wheat supplies by the Allies in the first two years of the war was a secret joint purchase effected by the governments of the United Kingdom, France, and Italy in the winter of 1915–16. *Ibid.*, p. 374.

[25]The average yield of wheat per acre in the prairie provinces in 1915 was 24 bushels, approximately 50 per cent above the long-time average, and a figure unequalled in western Canadian crop experience until 1952, when the average yield in the prairie provinces was 26.3 bushels per acre. See W. Sanford Evans Statistical Service, *Canadian Grain Trade Year Book* (Winnipeg, annually).

was fundamentally altered by the predominance in each of them of the single overseas buyer with unprecedented resources at its disposal. It was only a matter of time until formal alteration was made to accord with realities.

During the winter of 1916–17 the British government made an attempt to negotiate the purchase of wheat outside the newly established purchasing agency. In February, 1917, an offer was made to Hon. George Foster, Minister of Trade and Commerce for Canada, for the purchase of the entire Canadian crop at $1.30 per bushel for No. 1 Northern in store at Fort William and other grades in proportion. The Canadian Council of Agriculture advised the government that the offer was unsatisfactory, since the price had averaged more than that over the preceding six months, and it countered with a proposal to the government that the crop be offered at prices between a minimum of $1.50 and a maximum of $1.90, or a flat $1.70 per bushel. The British offer was repeated and no agreement was reached.

Circumstances sufficiently disruptive to induce the formal alteration of the open market system of trading in the Winnipeg Grain Exchange occurred for the first time in the spring of 1917. The crop of 1916 in the United States was small, and the Canadian crop, although fairly large,[26] was badly rusted and of poor quality. Much of it proved on delivery and inspection to be below contract grades, that is, below those grades deliverable at the seller's option on a futures contract.[27] The respective Wheat Export companies in the United States and Canada, acting as purchasing agencies for the British Royal Commission on Wheat Supplies, bought wheat and wheat futures heavily throughout the winter, the latter purchases being made without speculative intent but rather to assure the necessary supplies of wheat. By the spring of 1917 these agencies were in possession of great quantities of May and July wheat, that is, they had negotiated contracts to secure delivery of substantial quantities of wheat of any one of the contract grades during the months of May and July.

It became increasingly clear that the available supplies of wheat of contract quality were inadequate to meet the commitments held in good faith by the Allied purchasing agencies. An unpremeditated but none the less effective corner had been created in May wheat in the Winnipeg and Chicago markets. Although the purchase of futures by the purchasing

[26]At 262.8 million bushels the Canadian wheat crop of 1916 was above the ten-year average, 1911–20, of 238.4 million bushels. *Ibid.*, 1920–1, p. 4.

[27]Contract grades of wheat in Canada at that time were No. 1 Hard, and Nos. 1, 2, and 3 Northern.

agencies had, in effect, comprised hedging rather than speculative transactions, these agencies desired to obtain wheat of satisfactory milling quality and were by no means willing to waive their contractual rights by accepting grain of quality lower than the contract grades. Near panic developed in the Winnipeg market by April as "the shorts attempted to cover." During the month of April, May wheat rose in Winnipeg from $1.90 to over $3.00 per bushel.[28]

Faced with unmistakable proof that the futures market as constituted was inadequate to the circumstance of the day, and in order to avoid its complete collapse, the Council of the Winnipeg Grain Exchange made emergency arrangements on its own authority and requested intervention by the Dominion government for the longer run.[29] On April 28, 1917, the Council forbade trading in futures for speculative purposes and on May 3 the May and July futures were withdrawn from trading entirely. The Dominion government negotiated with the Wheat Export Company for a settlement of the impossible position of the short sellers of futures and the British purchasing agency accepted non-contract grades at a discount.

The Dominion government quickly made plans for the establishment of adequate control over the movement of Canadian grain, and the open market system of trading in the Winnipeg Grain Exchange came to an end for the duration of governmental control. On June 11, 1917, an Order in Council established the Board of Grain Supervisors, an agency of the federal government, and endowed it with monopoly control over Canadian wheat. The purposes of the Board, as expressed in the Order in Council, were to control the distribution of Canadian grain as between domestic requirements and the Allied purchasing agencies, and to regulate domestic distribution "in such manner and under such conditions as will prevent to the utmost possible extent any undue inflation or depreciation of values by speculation, by the hoarding of grain supplies, or by any other means."[30] The Board's principal powers were to acquire Canadian grain; to fix prices for it which would be uniform throughout the country with due regard for position, costs of transportation, quality, and grade; and to resell it to domestic millers and to Allied purchasing agencies. The Board handled no grains other than

[28]From April 1 to May 3, 1917. Sharp, "Allied Wheat Buying," p. 381.
[29]Ibid., pp. 381–2.
[30]As cited in *Report of the Royal Grain Inquiry Commission, 1938* (Ottawa: King's Printer, 1938), p. 31. Dr. Robert Magill, chairman of the Saskatchewan Elevator Commission of 1910 and first chief commissioner of the Board of Grain Commissioners under the Canada Grain Act of 1912, was appointed chairman of the Board of Grain Supervisors.

wheat because the purchasing agencies would not guarantee to take them at fixed prices.

The Board handled the remainder of the 1916 Canadian wheat crop and the crops of 1917 and 1918. The prices which it established were $2.40 per bushel, basis No. 1 Northern, Fort William, for the portion of the 1916 crop which remained unsold at the time it commenced operations, and, on the same basis, $2.21 for the 1917 crop and $2.24 for the 1918 crop. An order of the Board terminated trading in wheat futures in the Winnipeg Grain Exchange starting September 1, 1917. This order remained in effect until July 21, 1919.

It may be noted in passing that governmental control replaced the open market system in the United States as well as in Canada in the summer of 1917.[31] The United States Food Administration was established on August 10, 1917, and four days later the Food Administration Grain Corporation was created with power to buy wheat to support or guarantee a "fair price" to be set by a committee appointed by the President of the United States. The Corporation operated for the 1917 and 1918 crops, buying heavily in the autumn months to maintain the guaranteed price and selling its stocks throughout the remaining months of each year. At the end of June, 1919, the Food Administration Grain Corporation was dissolved, and replaced by the United States Grain Corporation,[32] which stood by to guarantee a floor price for wheat as had its predecessor. The basic guaranteed price for wheat in the United States throughout the three crop years was $2.20 per bushel. For the 1919–20 crop year the guarantee was largely inoperative due to the fact that market prices ordinarily ranged well above it.

The armistice of November 11, 1918, came at a time when the new crop year was well started. There was no question but that the Board of Grain Supervisors should continue to market Canadian wheat for the balance of the crop year. The policy to be pursued for succeeding years, however, remained to be decided. Joint Allied purchasing disappeared shortly after the armistice although governmental procurement and distribution of grain in most of the countries of western Europe persisted. There were influential groups in Canada actively interested in the resto-

---

[31]For a discussion of the wartime control of wheat in the United States see Frank M. Surface, *The Grain Trade during the World War* (New York, 1928); and Surface, *The Stabilization of the Price of Wheat during the War and Its Effect upon the Returns to the Producer* (Washington: United States Grain Corporation, 1925).

[32]The United States Grain Corporation came to an end May 31, 1920. It purchased over 138 million bushels of the 1919 crop in the course of its stabilization activities. See Surface, *The Stabilization of the Price of Wheat*, pp. 18, 24.

ration of the pre-1917 system of open market trading on the Winnipeg Grain Exchange. In general terms it was the view of the Dominion government that, with the war now over, pre-war principles and practices should be restored to effectiveness with a minimum of delay, and that the Canadian economy should return to "normal" as quickly as possible. There appeared considerable likelihood that the government would withdraw from the grain business and permit the "private trade" to operate as before.

The government permitted the reopening of futures trading on the Winnipeg Exchange on July 21, 1919, but the erratic speculative activity which followed indicated that the action was premature to say the least. Within the few days during which the futures market remained open, the price of October wheat advanced from $2.24½ to $2.45½.[33] By Order in Council of July 31 the Dominion government established the Canadian Wheat Board with the exclusive responsibility of handling and marketing the 1919 crop along with any part of the 1918 crop which had not been delivered to the Board of Grain Supervisors by August 15, 1919. In August, the Supreme Economic Council, organized by the Paris Peace Conference, restored the Royal Commisson on Wheat Supplies as the joint purchasing agency for Great Britain, France, and Italy. The purchase and distribution of wheat continued on a national basis in the other countries of northwestern Europe.

The constitution of the new Canadian Wheat Board is described in the records of the United Grain Growers as "almost identical with that submitted by the Canadian Council of Agriculture."[34] The plan was worked out by the United Grain Growers Limited and the Saskatchewan Cooperative Elevator Company and was submitted to the Dominion government on acceptance by the Council.[35] The original resolution of the Council had called for the establishment of a board "similar to the U.S. Grain Corporation with like power and functions and with the financial accommodation adequate to its operations."[36] The constitution as finally

[33]United Grain Growers Limited, *The Grain Growers Record, 1906 to 1943*, p. 25.

[34]*Ibid.*, p. 25.

[35]Joseph C. Mills, "A Study of the Canadian Council of Agriculture," unpublished M.A. thesis, University of Manitoba, 1949, p. 125.

[36]Resolution of the Canadian Council of Agriculture, July 9, 1919. The resolution contained the statement that: "The Canadian Council of Agriculture is strongly opposed to the opening of the Canadian markets for unrestricted trading in wheat, and would reiterate its recommendations of August 19, last year [1918], that the Government of Canada create, without delay, a body similar to the U.S. Grain Corporation with like power and functions and with the financial accommodation adequate to its operations." As cited in Mills, "A Study of the Canadian Council of Agriculture."

drafted and submitted to the government, however, must have departed from the resolution, for the resultant Canadian Wheat Board differed markedly in function from the United States Grain Corporation. The latter, as noted above, was purely a minimum-price support board and bought wheat only when necessary to that purpose. The Canadian Wheat Board was given a monopoly in the sale of the Canadian crop both in domestic and foreign markets and, in the latter markets, exclusive control of the sale of flour as well as wheat. It differed from its Canadian predecessor, however, in that it did not buy and sell at fixed prices but rather received all Canadian wheat at a fixed advance or initial payment and distributed the additional funds secured from its total sales in the form of interim and final payments. The total proceeds, grade by grade, were thus "pooled" as at Fort William[37] regardless of the actual proceeds of any individual consignment or of the time of delivery. This feature of the plan which, according to the Board, "resembled very closely that which was in existence in Australia,"[38] was of the utmost significance in its influence on agricultural agitation and organization during the 1920's.

The initial payment made by the Board was $2.15 per bushel, basis No. 1 Northern in store at Fort William. Along with this payment went participation certificates which entitled the seller of the wheat to a proportionate share in the total net proceeds of the crop. The initial payment was slightly below the price guaranteed to farmers in the United States ($2.25 per bushel) and farther still below the maximum to which prices had risen in the brief July interval of open market trading on the Winnipeg Grain Exchange. Since they had no experience in the disposal

[37]Freight from the local point to Fort William was deducted from the initial payment and was accordingly not pooled.

[38]The pooling idea was new to the Canadian West and had only recently been introduced for trial in Australia. The Board reported (*Report of the Canadian Wheat Board, Season 1920* (Ottawa: King's Printer, 1921), p. 6):

"In the first place, the board in adopting the plan of operation outlined in the Government's instructions, had to blaze a new trail. There was no precedent to follow. While a wheat 'pool' was being tried in Australia, its success had not been established, and it seemed to be regarded with more or less disfavour by some important sections of that country. In North America nothing of the kind had even been attempted. Some of the ablest men in the North American grain trade considered the plan as too 'communistic,' and doomed to failure. This impression was not confined to grain trade men alone, but was quite prevalent among our bankers and in other business circles. Large sections of the rural communities in the various provinces too, protested, by resolution or delegation, against the creation of the board, and as an alternative seemed bent upon having the Government either purchase the crop outright at a fixed price, or establish an organization similar to the United States Grain Corporation.

"In the rural districts along the international boundary, particularly in southern Manitoba and southeastern Saskatchewan, the cry during the autumn months of 1919 was for an open market. . . ."

of their crop on the basis of partial payment, and no certainty about the additional payment to be secured, the western growers were far from unanimously favourable to the new marketing arrangement. Some would have preferred to dispose of their grain at a higher fixed and final price, as they had been able to do under the Board of Grain Supervisors, some urged the adoption of the United States system of a guaranteed floor price at a higher level and without pooling, and many would have been willing to take their chances with the open market. Participation certificates were poorly regarded and were in many cases disposed of throughout the year for trifling sums. In May, 1920, the Board announced that the certificates would be redeemed for not less than 40 cents per bushel. An interim payment of 30 cents per bushel on July 15 and a final payment of 18 cents on October 30, 1920, brought the total payment for the compulsory government pool—for that, in effect, was what it was—to $2.63 per bushel, basis No. 1 Northern, Fort William.

Late in the session of 1920 the Dominion government passed an enabling act to permit continuance of the Wheat Board for the crop of 1920 should it be found necessary. Resumption of futures trading in the United States grain exchanges on July 15, 1920, and the concurrent relaxation of centralized buying, however, provided the government with the opportunity to announce that the operations of the Wheat Board would not be continued for the 1920 crop. The Wheat Board, in its report for the crop year 1919–20, pointed out that since it had been designedly a one-year board, it had used the facilities of the private grain trade as far as possible in order that "the trade would be better able to resume the handling of the wheat at the expiration of the controlled period."[39] By the early autumn of 1920 the "controlled period" was at an end.

[39]*Ibid.*, p. 5.

PART THREE

THE OPEN MARKET REJECTED

CHAPTER TEN

The Open Market System in the Inter-War Years

THE DOMINION GOVERNMENT had intervened in the grain trade during the First World War out of clearly demonstrated necessity rather than a change in its economic philosophy. It had, therefore, restored the *status quo ante bellum* at the first opportunity and extricated itself from its temporary marketing responsibilities by re-establishing the open market system for Canadian wheat and, in particular, by reinstating the Winnipeg Grain Exchange as the predominant institution of the Canadian grain trade. A drastic decline in the price of wheat followed.[1] The fact that all agricultural prices collapsed simultaneously if somewhat irregularly was not sufficient to modify the conviction which spread throughout the agricultural community that restoration of the open market system was largely responsible for the destruction of grain values. The price secured by the western farmer for his wheat under the Wheat Board had been the highest on record. The removal of the Board brought a return to the lower prices so familiar to those who had marketed grain through the open market before the war. The double coincidence was too striking to be overlooked. The Grain Exchange system was apparently responsible for, or at least inevitably associated with, low prices for wheat and would therefore have to be replaced. Agrarian protest before the war had been directed against what appeared to be a major abuse of the open market system, the domination of the Exchange by monopoly interests. After 1920 the protest was much more fundamental—the Exchange was regarded as detrimental to the growers' interests and would consequently have to be removed. The protest which developed in these circumstances and in accord with the interpretation which the growers placed upon them led to the establishment of the western wheat pools which will be

[1]The Winnipeg Grain Exchange was reopened to futures trading in wheat on August 18, 1920. The price of wheat in Winnipeg, basis No. 1 Northern, Fort William, for spot delivery, averaged $2.73½ for September, 1920; $2.31⅛ for October; $2.05 for November; $1.93½ for December; $1.94¼ for January, 1921; $1.88½ for February; $1.90⅞ for March; $1.76½ for April. From May to August, inclusive, the monthly average held steady above $1.80 per bushel. The average for September, 1921, however, was $1.48⅛, and for October, $1.15⅛. Sanford Evans Statistical Service, *Canadian Grain Trade Year Book, 1920–21*, p. 21.

dealt with in the next chapter. At this point it is necessary to outline briefly the general nature of what is commonly called the open market system of grain marketing and especially to consider the place of the Winnipeg and other commodity exchanges within such a system.

World production and trade in wheat in the 1920's differed in important respects from those of the years before 1914. World wheat production, which approximated 4½ billion bushels annually before 1914, had increased to approximately 5½ billion bushels annually in the late 1920's.[2] Only a small part of the total production entered into world trade channels and crossed international boundaries. For the five crop years 1909–10 to 1913–14 an average of 686 million bushels, or approximately 15 per cent of total production, was exported from the producing countries as wheat or wheat flour. Of the 5½ billion bushels produced annually during the years 1927 to 1931, 800 million bushels were exported each year, or again 15 per cent of the total.[3] During the 1930's, when world wheat exports fell to an annual figure of 550–600 million bushels, world trade scarcely exceeded 10 per cent of total world production.

Of greatest importance for the North American wheat economy in the inter-war years was the changed origin of the exports as compared with the years before 1914. Prior to the First World War Russia was the world's largest exporter of wheat, exporting annually an average of 164 million bushels or upwards of one-quarter of the world total in the five years from 1909–10 to 1913–14. After 1914, with the exception of two or three particular years,[4] Russia was no longer of significance in world wheat markets, and Canada, the United States, Argentina, and Australia emerged as the four major wheat exporters of the world.[5] With due allowance for annual variation and for lesser but by no means negligible wheat movements elsewhere, the bulk of the inter-war movement of export wheat converged from these four widely scattered areas upon Britain and western Europe.

At strategic points on the trade routes over which wheat moved in international commerce, market centres had been established and were

[2]Statistical Appendix to C. F. Wilson, "An Appraisal of the World Wheat Situation" in *Proceedings of the Conference on Markets for Western Farm Products* (Winnipeg, 1938), Tables 1, 11, and 12, pp. 27, 45.

[3]*Ibid.*, Table 15, p. 50.

[4]Maximum exports of Russian wheat in any of the inter-war years were the following: 49 million bushels in 1926–7; 112.6 millions in 1930–1; and 63 millions in 1931–2. *Ibid.*, Table 11, p. 45.

[5]Major exporters are not, of course, to be confused with major producers. The four major wheat *producers* in the inter-war years, in descending order of acreage were Russia, the United States, China, and Canada. *Ibid.*, Tables 1, 11, and 12.

well developed and integrated by 1920. Traders congregating in these centres had evolved elaborate institutions for facilitating and regulating their activities. A detailed listing of these institutions is beyond the scope and purpose of this study. With particular reference to exporting countries, however, the grain markets at Winnipeg, Chicago, Buenos Aires, and Rosario (Argentina) were of special international importance by the end of the First World War. Of the many market organizations in import countries, those at London and Liverpool had been prominent over a long period of time, the London market as the leading purchaser of grain on British and European account, and the Liverpool market as a purchaser of grain but more particularly as the central futures market of the world. During the inter-war years, Liverpool, Chicago, Winnipeg, and Buenos Aires formed a closely integrated group as the leading futures markets in the world grain trade.[6]

Geographic location has given to Winnipeg a measure of importance in the Canadian grain trade much greater than that which might be suggested by the population, industry, or grain consumption of the city or, indeed, by the quantity of grain produced within the boundaries of the province of Manitoba. The transcontinental railways, with their vast network of feeder lines covering the prairie wheat growing area, converge upon Winnipeg as on a port of exit from the West. Beyond the city, toward the East, the lines again diverge, some running to the head of the Lakes at Fort William–Port Arthur, some running directly overland to the eastern seaports of Montreal and Quebec, and some running south to the United States. Before the opening of the Panama Canal, effective in the 1920's, all wheat exported from the prairie provinces, whether to Britain and Europe or to the eastern Canadian market, passed through Winnipeg. After that time, significant quantities of prairie wheat moved westward through Vancouver and Prince Rupert directed either toward markets in the Orient or toward European markets via the Panama Canal. More recently, small quantities of wheat have been exported via the Hudson Bay Railway and Port Churchill. Nevertheless, Winnipeg continues to be the central market for Canadian wheat and the exclusive Canadian market with facilities for trading in grain futures.

Attempts to organize a grain exchange in Winnipeg date from the years when the Canadian Pacific Railway first linked the West with eastern markets and Atlantic ports. An unsuccessful move toward organization in 1883 was followed by efforts which led to the formal establishment of the Winnipeg Grain and Produce Exchange in 1887. This

[6]See articles, "The Grain Markets of Britain," Midland Bank Limited, *Monthly Review*, July–Aug., 1930, pp. 4–7, and Aug.–Sept., 1930, pp. 5–8.

association was incorporated in 1891 by an act of the Manitoba Legislature.[7] Since there was no provision for futures trading during the early years of operation of the Exchange, it was necessary for exporters to buy grain outright and to bear the risk of price changes until their purchases were disposed of at an Atlantic port, generally New York.[8] The lack of hedging facilities, combined with many inadequacies in the matter of physical supply, conspired to create wide margins in the price structure. One alternative used to a certain extent was to hedge Winnipeg purchases by selling futures in Chicago and in other organized futures markets. The cumbersome nature of this process, however, indicated a need for the establishment of local facilities. The Winnipeg Grain and Produce Exchange Clearing Association was consequently established and incorporated under the Joint Stock Companies Act of Manitoba in June, 1901. After a considerable delay the organization was finally completed, and on February 2, 1904, futures trading and the clearing of futures trades began in the Winnipeg market.[9] The events which led to the reorganization of the Exchange as a voluntary, unincorporated association in 1908 were outlined in chapter VIII.[10]

The Winnipeg Grain and Produce Exchange and its counterpart, the Winnipeg Grain and Produce Exchange Clearing Association, are designated collectively in common terminology as the Winnipeg Grain Exchange. Since 1904 they have constituted the central institution and the symbol of the "open market" or "competitive" system for the marketing of Canadian grains. The essential distinguishing feature of this system is its provision for the transfer of title to cereal products by dealers working within a highly organized institutional framework which embraces a futures market and which operates with a minimum of external control whether on the part of government, growers, or consumers.

Although various circumstances may constitute an obvious modification in the open market system, such a system is most clearly rendered inoperative by governmental decree which bans futures trading and which at the same time gives to government a monopoly of the marketing of a Canadian grain or grains. It is clear, therefore, that the open market system may persist for some grains while it has been rendered

[7]An Act to Incorporate the Winnipeg Grain and Produce Exchange, *Statutes of Manitoba*, 54 Vic., c. 31 (1891). See *Report of the Royal Grain Inquiry Commission*, 1925 (Ottawa, 1925), pp. 121–2.

[8]*Report of the Royal Grain Inquiry Commission*, 1938 (Ottawa, 1938), pp. 28–9.

[9]See evidence of Frank O. Fowler, manager, Winnipeg Grain and Produce Clearing Association, before the Special Committee of the House of Commons on the Marketing of Wheat and Other Grains under Guarantee by the Dominion Government, *Minutes of Proceedings and Evidence* (Ottawa, 1936), p. 27 ff.

[10]See pp. 137–8.

inoperative for others. Wheat has on occasion been removed from open market channels in Canada to be placed under the exclusive control of a governmental agency, while coarse grains and flax have been left to the open market. It is also evident that governmental or growers' activity may provide the grower with alternatives to the open market system.[11] Thus he might be permitted to choose, for example, between disposing of his grain through the agencies of the open market or through a government marketing board, or between the agencies of the open market and a pool operated by growers. All these possibilities have been explored within the past half-century of Canadian grain marketing experience.

Although the constitution of the Winnipeg Grain Exchange has remained essentially unchanged from 1908 to the present time, the institution can be said to have operated continuously only if account be taken of a variety of functional limitations to which it has occasionally been subjected. The continuity, in fact, relates only to grains other than wheat, for throughout two protracted intervals there has been neither trading in wheat futures nor open market trading in wheat in the Winnipeg market. The first such interval, described in chapter IX, continued from September 1, 1917,[12] to August 18, 1920, with a ten-day break in 1919. The second interval began on September 27, 1943.[13]

Throughout the 1920's and 1930's the Winnipeg cash and futures wheat markets operated freely but with substantial modifications in the open market system as compared with preceding decades. The pools which were organized in 1923 and 1924 rapidly developed to the point where they handled approximately one-half of all the wheat produced in the prairie provinces. They acquired privileges of membership in the Winnipeg Grain Exchange but made it a point to sell directly to overseas buyers through agencies which they established and maintained abroad. During the early 1930's the pool marketing of wheat was replaced by a federal stabilization organization. From 1935 to 1943 the newly established Wheat Board provided an alternative channel. These developments are described in later chapters. They are mentioned here only to indicate that the open market system of the inter-war years was not the exclusive grain handling medium in Canada that it had been before the First World War.

[11]See chap. XI, pp. 198–9.

[12]The Winnipeg Grain Exchange had already voluntarily suspended futures trading after an investigation, in May, 1917, into speculative price fluctuations. President's address, Winnipeg Grain Exchange, *Ninth Annual Report*, Sept., 1917, as referred to by H. S. Patton, *Grain Growers' Coöperation in Western Canada* (Cambridge, Mass., 1928), p. 195 n.

[13]For an outline of the action taken at that time, see chap. XIV, pp. 275 ff.

The Winnipeg Grain Exchange differs in no fundamental respects from organized commodity exchanges in general. It does not buy, sell, or handle grains. Its essential purposes are to provide a meeting place for those who do buy or sell or who assist in any way in the marketing of Canadian grains, to establish and ensure uniformity of trading practices among its members, and to assemble and make available to its members that intangible but all-important body of diverse information collectively known as market news.[14] Only members may trade on the Exchange, either on their own account or for others. The latter transactions are carried on for commission at rates set and controlled by the governing body of the Exchange. In the early 1920's the membership of the Winnipeg Exchange totalled 355, of whom 267 were engaged in the handling and marketing of cash grain and 50 were primarily engaged in futures trading.[15] The group concerned with cash trading included the representatives of elevator companies, cash grain commission merchants and brokers, millers and maltsters, shippers and exporters.

Cash grain contracts may be entered into in the Winnipeg Grain Exchange for "spot" grain, "billed and inspected" grain, or "on track" grain, depending on the position of the product. Spot grain is in store in a regular terminal elevator at Port Arthur or Fort William, and its title passes by surrender of the warehouse receipt. Billed and inspected grain is in transit and has passed through an inspection point. It has therefore

[14]The statement of purpose in the constitution (1908) of the Winnipeg Grain Exchange (as quoted in the *Report of the Royal Grain Inquiry Commission, 1925*, p. 122), is relevant and places primary emphasis on market news:

"(a) To compile, record and publish statistics, and acquire and distribute information respecting the grain, produce and provision trades, and to promote the establishment and maintenance of uniformity in the business customs, and regulations among the persons engaged in the said trades; to inaugurate just and equitable principles in trade, and generally to secure to its members the benefits of legitimate co-operation in the furtherance of their business and pursuits.

"(b) To organize, establish and maintain an association, not for pecuniary profit or gain, but for the purpose of promoting objects and measures for the advancement of trade and commerce respecting the grain, produce and provisions trades for the general benefit of the Dominion of Canada, as herein provided; to acquire, lease or provide and regulate a suitable room and place for a Grain and Produce Exchange and offices . . .; to facilitate the buying and selling . . .; to avoid and amicably adjust, settle and determine controversies and misunderstandings between persons engaged in the said trades. . . ."

[15]*Ibid.* The membership as described in the early 1940's included: ". . . shippers and exporters, millers and other processors, grain and feed merchants, owners of country elevators . . . terminal elevators . . . and mill elevators, farmers' organizations, wheat pools, banks, railway companies, vessel owners, commission merchants, brokers, and foreign grain concerns located in almost every corner of the globe." See George S. Mathieson, *Wheat and the Futures Market: A Study of the Winnipeg Grain Exchange* (Winnipeg, 1942), p. 5.

received official government grading. "On track" grain is loaded in rail-way cars. It is, to that extent, in transit, but it has not yet received official inspection. Bills of lading properly endorsed, and railway advice notes, effect transfer of title to the latter types of grain.[16]

The functioning of the cash market segment of commodity exchanges is readily understood and is seldom seriously criticized. The purposes and functions of futures markets are, however, not so easily explained, they may readily be misconstrued, and they frequently fall under bitter public attack. The major conflicts which developed in relation to the Canadian grain trade after the First World War and the evolution of policy concerning the marketing of Canadian grains throughout the past thirty-five years are understandable only in reference to the activities of futures markets as exemplified in the Winnipeg Grain Exchange. It would be useless to attempt a portrayal of grain marketing policy in Canada in the inter-war years without a brief consideration of the nature of organized futures trading.

The Winnipeg grain futures market, like any futures market, is an organization for buying and selling futures, and these, in turn, are highly standardized contracts for the future delivery of a specific commodity or commodities. There is nothing unusual in the existence of contracts specifying the conditions under which goods or services are to be delivered at some future time. Future dealing, future trading, or contracting, is an indispensable commonplace in the economic system, so much a commonplace in certain segments of the economy that a wide variety of persons, particularly in the construction industry for example, are known as "contractors." It is frequently argued that futures markets are designed solely to assist in the negotiation of pledges of future performance in certain areas of economic life and that, since commitments regarding the future are unavoidable, there is particular merit and no possible hazard in having special institutions to facilitate their negotiation.

Despite many similarities, there are essential differences between organized futures trading and the wide range of ordinary business engagements providing for future performance. The high degree of stan-

---

[16]*Report of the Royal Grain Inquiry Commission*, 1925, p. 124. When speaking of grain marketing in Canada it is customary to speak of the "Winnipeg price" of wheat or other cereals. The term refers to the price which is in effect at any point of time in the Winnipeg Grain Exchange, and does not indicate the location of the grain. Winnipeg prices may be cash prices or futures prices; the cash prices may be spot, or billed and inspected. If any single price deserves more than the others to be regarded as *the* Winnipeg price of wheat it is the spot price, that is, the price of wheat in store in a regular terminal elevator at Fort William or Port Arthur and available for immediate delivery on the surrender of warehouse receipts. Fort William is approximately 400 miles from Winnipeg.

dardization imparted to contracts by the rules and regulations of commodity exchanges gives them a measure of liquidity rarely if ever attaching to the commitments of ordinary business usage where each contract is specific to a specific situation. People *enter into* contracts for the construction of houses, the delivery of autos or household equipment, or the employment of labour; they *buy* or *sell* contracts, or futures, providing for the delivery of products on organized futures markets. The ordinary expectation concerning contracts for future performance which are entered into outside of futures markets is that each contract will mature and eventuate in the performance provided for in its terms. The house will be constructed, the merchandise will be delivered, the workers will work for the specified wages. In the futures market, however, contracts are not exclusively or even normally fulfilled by making and taking delivery as contracted for. Fulfilment, which may, of course, be accomplished by the transfer of warehouse receipts for grain in store, is more commonly achieved by a process of offsetting, the vendor offsetting his sale by a subsequent purchase and the purchaser his purchase by a subsequent sale.[17]

The elements of standardization provided by the futures contract and by the rules and regulations of the exchange governing such contracts may be identified under the following headings: (1) the commodity, (2) the quantity, (3) the range of quality within which delivery is permissible, (4) the month of delivery, (5) the nature of the option concerning specific grade and date of delivery, that is, whether it is a seller's or a buyer's option, and, finally, (7) the price. Detailed elaboration of these points is unnecessary. Taken collectively they readily explain the high degree of liquidity of futures contracts and constitute the conditions under which such contracts continuously bear a market price and are bought and sold with a freedom achieved by few commodities.

The vigorous agrarian opposition which developed in the early interwar years against the Winnipeg Grain Exchange was concentrated primarily if not exclusively on the futures market and on transactions

[17]A student of futures trading on American commodity exchanges states that in the middle 1920's approximately one-third of 1 per cent of the total futures trades in wheat, corn, oats, and rye in the United States were closed out by delivery of warehouse receipts, that is, by delivery. Corresponding estimates of the Winnipeg market are not available and the proportions may differ considerably from those suggested by American experience. The fulfilment of futures contracts by the delivery of the commodity is nevertheless only one of the normal methods of fulfilment in futures markets. See G. Wright Hoffman, *Futures Trading upon Organized Commodity Markets in the United States* (Philadelphia, 1932), pp. 106–9.

involving futures contracts. Not all such transactions were suspect, but the open market system apparently required complete freedom for the purchase and sale of futures without regard for any distinction between those which were clearly beneficial and those which might be detrimental to the producers. Supporters of the open market system argued that futures trading was indispensable to hedging and that hedging, in turn, was one of the essential conditions for the extension of bank credit to elevator and milling companies and to the export trade. This line of reasoning was understandable and subject to little criticism. But futures were admittedly bought and sold for speculation as well as for hedging and it was the use of the futures market for speculative purposes that drew the concentrated and sustained fire of agrarian critics in the period after 1920. This point requires elaboration.

When the open market was freely operative in Canada, the speculator bought and sold futures on the Winnipeg Grain Exchange for the purpose of making a profit on price fluctuations. For his objectives the futures contract took on the characteristics of a commodity in its own right. It varied in price, and was bought, held, and sold again like any durable commodity. The speculator in the grain futures market, in so far as his speculative activities were concerned, had no grain nor any need or desire to have grain. He dealt exclusively in futures contracts. He bought a quantity of May wheat futures, for example, and thus entered into a contract to accept delivery of a specific quantity of wheat within the month of May next. He had no intention of accepting delivery then or at any time but bought the future on the strength of a belief that it would rise in price at some time prior to the future month and could then be sold at a profit.[18] Or again, the speculator *sold* a quantity of May wheat futures as an initial speculative transaction. He thereby pledged himself to make delivery of that quantity of wheat within the month of May. He had no wheat nor any intention of making delivery within the

[18]Actual contracts, once bought, were, of course, not resold. A speculator who bought 10,000 bushels of May wheat from another speculator or from a hedger registered that purchase in the clearing-house and was thereupon obligated to conduct any further dealings concerning the contract with the clearing-house. Rather than accept delivery from the clearing-house of title to 10,000 bushels of wheat in the form of warehouse receipts—and this, as a speculator, he would have no intention of doing—he could cancel his obligation to accept delivery at any time by *selling* 10,000 bushels of May wheat to another member of the Exchange, either a speculator or a hedger. On registration of the second transaction with the clearing-house, the speculator's obligations were cancelled, and on the immediate cash payment of the settlement, whether by the speculator to the clearing-house or by the clearing-house to the speculator, as the case might be, the entire speculation was concluded.

future month. He sold the future in the belief that its price would fall at some time before the end of the future month and that it could then be bought in at a profit.[19]

These were among the most important of the observed circumstances that led to the dissatisfaction of Canadian wheat growers with the Winnipeg Grain Exchange after the First World War. The central criticism which emerged then and which western wheat growers would argue as fully valid to the present day is that the price received by the producer for his grain when disposed of through open market channels is largely at the mercy of the speculator and of speculative activity. It had long been observed, and as the controversy developed after 1920 it was noted as having particular relevance, that the cash prices for wheat—in Canada the prices for wheat in spot, track, and street positions—bore a close relation to the prices of wheat futures. It was noted further that dealings in futures for hedging purposes were dependent upon actual movements of grain, that hedging sales were heavy at times when marketings of grain were heavy and that hedging purchases of futures were substantial when milling and export demand was heavy. The supply of and demand for wheat futures for hedging purposes were therefore dependent on the supply of and demand for wheat. It was observed finally, however, that the supply of and demand for futures for speculative purposes were independent of wheat movements and of the hedging requirements of grain handlers and processors, and were, in short, determined exclusively by the decisions of speculators whose sole concern was the maximization of speculative profit. Since cash and futures prices were closely interrelated and since speculators were able to alter the supply and demand of futures at will, it is readily understandable that the speculator should have been endowed in the producer's mind with substantial responsibility for the open market price of wheat and other grains.

Western growers complained persistently in the inter-war years about the general influence of the speculators on prices, and maintained specifically that the speculators' influence was seasonal, and therefore particularly detrimental to the interests of the producer. In the months immediately after harvest, the concentration of Canadian grain on the market and the consequent heavy offerings of hedging futures appeared to create a perfect opportunity for the speculator to profit at the expense of the producer. If there were no speculative support in the futures market during these months, the prices of futures, and the price of wheat as well, would obviously decline severely. Under these circumstances

[19]The explanation given in note 18 applies to the present case, *mutatis mutandis*.

the speculative groups would appear to have nothing to gain from buying futures in amounts adequate to maintain the market and, indeed, much to gain from refusing to do so. The farther down the various prices were permitted to sag under hedging pressure, the cheaper both cash wheat and futures would be to purchase and the more certain it would be that they could be disposed of later in the crop year at an attractive speculative profit.

Furthermore, it was argued, passive refusal to support the market was not the only possibility open to the speculator. The open market permitted "short" sales, the sale of grain for future delivery by persons who had no grain and no intention of making delivery. The speculator did not need to remain idly by to watch wheat prices fall under the weight of hedging pressure but would have every incentive and opportunity to add to that pressure by selling short, by contracting to deliver in the future month grain which did not exist. Growers have never been convinced that this "phantom wheat," these "wind bushels," freely at the disposal of the futures speculator were not used effectively as a means of depressing the price of wheat and other grains to unreasonably low levels in the autumn months, during which western farmers have of necessity sold the bulk of their single annual product.

On this subject, there are two points of a general nature which are scarcely open to argument. First, a seasonal disparity in prices is normal for any commodity which is produced seasonally and consumed throughout the year. The process of holding a product over a period of time, the "creation of time utility," is an economic process to which cost inevitably attaches. The product must be stored, it may be insured, and throughout the entire time that elapses between its production and its consumption some one must finance its entire capital value. Cost is the inescapable counterpart of all these services. A bushel of wheat which is put on the market in the fall and not consumed till some months later is not the same economic quantity throughout this interval of time but is rather a unit to which additional costs of production constantly accrue. Other things being equal, the price should be lower immediately after harvest than at any later date, the normal difference in price at any two points of time being equal to the carrying charges involved. The existence of a seasonal disparity in price is not, by itself, a proof of market manipulation.

A second point to be made is that speculation is unavoidable whenever a commodity must be held, and when, during the same period of time, price movements are unpredictable. The only questions that arise, really, are who shall do the speculating, and under what conditions?

The advocates of the open market system have always contended that the organized futures market provides the obvious answer, since such a market creates the opportunity to transfer the risk to the deliberate speculator who accepts it in the hope of deriving a profit. The advocates of the pools spoke of doing away with the speculator and even of doing away with speculation. The pools did, in fact, bypass the speculator and the organized speculative market, but their method of handling grain made it necessary for their members to accept and pool the risk collectively. Advocates of the marketing of wheat by a monopolistic agency such as the Canadian Wheat Board envisage, in effect, the transfer of a maximum proportion of market risk to the federal government by the establishment of a high initial payment for the product, and the assumption by the growers, collectively, of the residual risk.

Three federal royal commissions[20] examined futures trading in the Canadian grain trade in the inter-war years with particular reference to the complaints of western growers. The pre-war view that the Winnipeg Grain Exchange was essentially beneficial, or at least innocuous, if it could be purged of monopolistic elements, had given way to the judgment that as an institution it was inherently and incorrigibly detrimental to the interests of western farmers. By 1923, when the first of the inter-war royal commissions faced the problem, this judgment had become crystallized and unequivocal. The Commission of necessity considered at length the protests which emerged profusely from this belief and presented a detailed analysis of its views on the effects of futures trading. This analysis was not essentially modified or enlarged upon by either of the later federal commissions of the inter-war years except for a brief study appended to the report of the Commission of 1936–8.

The Commission of 1923–5 noted significantly that "the farmers' complaints against present methods of marketing grain focus upon the Winnipeg Grain Exchange as the head and shoulders of the present

[20]Hon. W. F. A. Turgeon was appointed chairman of a commission in 1923 "to investigate and report upon the subject of handling and marketing grain in Canada." See *Report of the Royal Grain Inquiry Commission*, 1925. Sir Josiah Stamp was chairman of a commission in 1931 directed to consider the effect of the dealing in grain futures upon the price received by the producer. See *Report of the Commission to Enquire into Trading in Grain Futures*, 1931 (Ottawa, 1931). Mr. Justice Turgeon was appointed chairman of a second federal grain inquiry commission in 1936 with instructions "to enquire into and report upon the subject of the production, buying, selling, holding, storing, transporting and exporting of Canadian Grains and Grain Products." *Report of the Royal Grain Inquiry Commission*, 1938. A commission appointed by the Meighen government in 1921 to conduct a general inquiry into the Canadian grain trade was stopped by an injunction secured by the United Grain Growers and upheld by the Supreme Court of Canada.

system."[21] Among the grievances expressed directly against the exchange itself the most important were the following: "That speculation either on the cash or futures market injuriously affects the farmer and the community: (a) the price of grain is thereby unduly depressed in the autumn when the farmers are selling the bulk of their crop; (b) lucrative profits are made by speculators, scalpers, etc., through gambling with the farmers' product; (c) disastrous losses are made in speculation."[22] The Commission dealt with this composite charge at considerable length.[23] It drew the standard distinction between hedging and speculation and indicated the importance of hedging to the various grain trading interests operating under open market conditions. It pointed out that the speculator was essential to assure a continuous market. It analysed the effects on prices of the activities of well-informed "professional" speculators operating under competitive conditions. It noted that speculators were not always well informed and that they therefore "did not always form the correct estimate of the future conditions of supply and demand."[24] Under these conditions, the Commission added, "society, far from benefiting from their activity, suffers. Fluctuations in price are magnified rather than reduced. . . ." Despite this qualification the Commission stated that in its view, speculative activity, in addition to being essential to hedging operations, tended to stabilize prices and reduce fluctuations therein, particularly from season to season. The conclusions of the Commission of 1923–5 regarding futures trading and a futures market are so pertinent to both agrarian protest and the development of federal wheat marketing policy in the inter-war years that quotation in full is justified:

1. That a futures market permits hedging and that hedging by dividing and eliminating risks in price variations reduces the spread between the prices paid to the farmer for his product and those obtained for it upon the ultimate market.

2. That hedging facilitates the extension of credit and thereby reduces the cost of handling grain by making it possible for grain dealers to operate on less capital than would be the case otherwise.

3. That for the same reason hedging makes a larger degree of competition possible in the grain trade, on a given amount of capital.

4. That hedging is of advantage to exporters so that even in instances where grain is handled under a pooling organization where the initial risk is carried by the farmer himself, in order to handle successfully the export trade such organizations find it desirable to make use of the futures market.

[21]*Report of the Royal Grain Inquiry Commission*, 1925, p. 128.
[22]*Ibid.*
[23]*Ibid.*, pp. 130–9.
[24]*Ibid.*, p. 133.

5. That a competent speculative element in the market ensures a continuous and searching study of all the conditions of supply and demand affecting market prices.

6. That speculative transactions tend to keep prices as between the contract grades and as between present cash prices and cash prices in the future in proper adjustment to each other and to future conditions of supply and demand.

7. That prices thereby tend to be stabilized and fluctuations reduced.

8. That a speculative element is necessary in an exchange to ensure a continuous market so that when a crop is dumped upon the market in the fall the farmer will not suffer loss by a heavy drop through absence of demand for immediate use.

9. That individuals who engage in speculative transactions without adequate knowledge or capital not only usually lose heavily but also are a disturbing element upon the market. Their transactions become mere gambling.

10. That it does not seem possible to legislate effectively so as to eliminate such individuals without disturbing the general and genuine usefulness of the exchange; but that legislation should be directed towards preventing the incompetent from being lured into speculation.

11. That Parliament should not at present enact restrictive legislation in the expectation of tempering fluctuations on the exchange, or of improving and stabilizing prices, but that time should first be taken to allow the new American law on this subject to demonstrate its efficacy.

12. That the penalties and precautions against rigging the market, or dishonourable trading, seem calculated to make such practices rare and unprofitable.[25]

The analysis and conclusions of the 1923–5 Commission confirmed existing official views regarding futures trading and substantiated existing marketing policies without even minor modification. The open market system was given a clean bill of health and emerged as the only sensible means of disposing of the Canadian wheat crop. The approach of the Commission to the problem of futures trading might be described as the textbook approach. Farmers' protests were recorded directly from the evidence but the refutation of these protests was constructed in theoretical terms. The importance of hedging in the open market system, as emphasized by the Commission, is unquestioned. The contribution rendered to the stability and continuity of a market by the activities of a "competent" speculative element is readily demonstrable in theoretical terms. The Commission recognized that some of the speculators were incompetent, and that they constituted a disturbing influence on the market, but it made no attempt to assess the relative importance of this disturbance. In its view little could be done to avoid it and legislative restrictions were likely to do more harm than good.

[25]*Ibid.*, p. 139.

The primary significance of the Commission of 1923–5 was that it reflected a changed attitude among western wheat growers vis-à-vis the government, in direct contrast to the attitude reflected by the commissions of 1899 and 1906.[26] The turn of the century had brought the first clear proof that the West could and would develop according to the pattern envisaged in the national policy. All that remained in doubt was whether any continuing impediments could be recognized and removed in time to prevent serious retardation of the rate of expansion. Marketing and transportation agencies were prepared to develop the West at their own pace and under conditions of monopoly, but it was those very conditions that constituted the greatest single impediment to the realization of the goals of the national policy after 1900. This fact was by no means universally recognized (the western grain growers saw it, of course) and was only gradually accepted by the Dominion government. Finally, however, the government did become convinced that monopoly would stifle rather than promote western expansion, and thus imperil the national policy. It found the strongest possible support for this belief in western agrarian protest, and it used the commissions of 1899 and 1906 chiefly for the purpose of getting this protest on the record and for educating the public and Parliament to the need for curbing monopoly in the western grain trade. So sure was the Dominion government of what it wanted to be forced to do, and so certain was it that on this point the views of western agrarian leaders were fully in accord with its own, that it manned its early commissions either exclusively or predominantly with farmers. The government was able to implement the commissions' recommendations promptly, because the views of the farmers who manned the commissions did not constitute a challenge to the government's basic philosophy.

The situation after 1920 was in sharp contrast to that described in the previous paragraph. There was no longer any possibility of harmony between the views underlying agrarian protest and those of any substantial section of federal leadership. The conviction of the western wheat grower that the Winnipeg Grain Exchange ought to be abolished because of inherent unsoundness was, even on the surface, a radical conviction. Far down it rested upon a belief—diametrically opposed to the free enterprise tenets underlying the national policy—that the open market or competitive system, the system of freely moving prices, ought not to govern the marketing of western grain. Complaints against existing methods of marketing grain focused, as the Commission of 1923–5

[26]For a detailed account of these commissions, see chap. IX, pp. 154 ff.

said, on the Winnipeg Grain Exchange "as the head and shoulders of the system." It was not, however, a question of abolishing the Exchange and leaving the grain trade otherwise intact. The proposal was that the free enterprise grain trade should be removed, and replaced by a governmental agency with power to control prices. The western wheat farmer reasoned that there was serious inadequacy in national policies which assured the subsidization of transportation interests by land and money grants and security guarantees, provided for the protection of industry by tariff walls, permitted to economic interests generally the greatest possible freedom in their efforts to avoid the hazards of competition by combination and agreement, and, at the same time, left the agricultural producer exposed to the full rigours of competition both nationally and internationally.

It is readily understandable that the Dominion government should select the personnel for the investigation of the new agrarian protest of the inter-war years with extreme care and with an eye for the personal qualification which might be described as politico-economic "soundness." With the appointment of the Commission of 1923–5 the western wheat grower was removed from the bench and was cast in the single unequivocal role of witness for the prosecution. The bench now held a judge, an economist, a railway official, and, with a formal bow to the agricultural interests at stake, the Dean of Agriculture of the University of Saskatchewan. A thorough grounding in the theory of nineteenth-century liberalism would be the best possible preparation for an evaluation of the claims of the western farmer in such a way as to leave freedom of enterprise intact. The Commission of 1923–5 avoided the really knotty problem implicit in the protests of the wheat growers of the 1920's by ignoring it. Wheat growers urged that the Grain Exchange be abolished, and the open market with it. The abolition of the system of freely moving prices for wheat and other western grains was inevitably implicit, but by no means clearly explicit, in this request. The Commission ignored the implied part of the request and tacitly assumed the continuance of the free price or supply and demand system of the marketing of wheat. Starting from this assumption, which it did not trouble to state, the Commission proceeded to prove the efficacy if not the indispensability of futures trading under the assumed circumstance.

The grain trade commissions of 1931 and 1936 followed the lead of the Commission of 1923–5 in their approach to the question of futures trading. The Stamp Commission of 1931 analysed futures markets in great detail and with an impressive display of abstract erudition. In conclusion, however, it quoted with approval the findings of the 1923–5

Commission and added little thereto. Its final word on the matter was as follows: "In brief, our answer to the question submitted is that in addition to the benefits reflected to the producer in furnishing a system of insurance for the handling of his grain, and in providing an ever-ready and convenient means for marketing the same, futures trading, even with its disadvantages of numerous minor price fluctuations, is of distinct benefit to the producer in the price which he receives."[27]

When Mr. Justice Turgeon returned to investigate the Canadian grain trade in the years 1936–8 he found the implications of the demand for the abolition of the futures market much more fully recognized and clearly stated in the submissions made by western farmers. "The demand made before me," said Mr. Turgeon, "for the abolition of futures trading in Canada was always accompanied by a further demand for the creation by the Government of a permanent National Wheat Board, whose duty it would be to take over the whole of our wheat crop each year and to market it both at home and abroad."[28] The request for a compulsory wheat board was so persistently made that he could not ignore it: "This proposal of a compulsory Government Board, asked for by nearly all the farmers' organizations, and by a great many of the individual farmers, who appeared before me, has preoccupied me more than anything else since the beginning of this inquiry."[29] His analysis of the proposal led him to reject it as a long-run possibility because of the dangers of political interference and the hostility of the private trade at home and abroad. The Wheat Board which had been established by legislation in 1935 was still in operation on the basis of voluntary patronage, and the commissioner was unable, in view of the condition of the market in 1938, to recommend its immediate dissolution. "Under what may be called normal conditions," he said, "open markets in the United Kingdom, a fair relationship between world supply and import demand, and no danger clouds on the immediate horizon, the Government should remain out of the grain trade, and our wheat should be marketed by means of the futures market (under proper supervision), and encouragement given to the creation of co-operative marketing associations, or Pools."[30] Conditions in 1938 were clearly not "normal"; hence the reluctant recommendation that the Wheat Board should not immediately be disbanded.

[27]*Report of the Commission to Enquire into Trading in Grain Futures*, 1931, p. 72. The "question submitted" was "what effect, if any, the dealing in grain futures has upon the price received by producers?"
[28]*Report of the Royal Grain Inquiry Commission*, 1938, p. 184.
[29]*Ibid.*, p. 185.
[30]*Ibid.*, p. 189.

The most specific of the complaints against the open market system in the middle thirties, as in earlier years, was that the futures market permitted or caused an excessive slump in grain prices in the autumn months when the bulk of the western crop is marketed. Mr. Justice Turgeon noted the complaint but made no comment of significance on the question of seasonal changes in price in the main body of his report. A study incorporated in an appendix,[31] however, came to grips with the question in limited scope but in a manner well designed to inspire confidence in its results. The question dealt with in the appendix was the basic one, to what extent were autumn prices for wheat unduly depressed in relation to the prices prevailing throughout the remainder of the crop year? The approach adopted was statistical and involved a series of comparisons of seasonal wheat prices in the Winnipeg market over the fifteen-year period from 1922–3 to 1936–7. The details of the analysis are beyond the scope of this study, but the conclusions are worth quoting in full:

1. There is an autumn decline, in relation to the previous summer, in both cash and futures prices, and a co-related rise which occurs chiefly in the following May or July, and sometimes in both these months.
2. Considered by themselves, cash prices do not indicate variations over the year as a whole greater than would be expected to be caused by mounting carrying charges, although the rise in prices which sometimes occurs between May and July indicates that other influences are also at work.
3. The course of futures prices, however, gives evidence of a tendency towards at least one and sometimes two periods of pronounced speculative price rises, almost always in May and/or July, and this speculative influence also accounts for a part of the rise in cash prices.
4. The decline in all prices in the autumn appears to be chiefly a recession from previous rises; but in the case of cash prices, at least part of such recession is natural in view of the change from old crop to new crop.[32]

These conclusions and their supporting analysis do not provide categorical proof that seasonal variations in wheat prices under open market conditions are not affected by speculative activity. They establish a

[31]*Ibid.*, Appendix VI, pp. 245–55.
[32]*Ibid.*, pp. 254–5. Professor Filley recorded in 1929 the results of certain comparisons of seasonal prices relating to wheat in the United States. Data for the average monthly prices for cash wheat at Chicago for forty-eight crop years, from July, 1880, to June 30, 1927, indicated an average annual spread of 10 cents per bushel, an amount "less than the cost of holding wheat for ten months in a public warehouse." H. C. Filley, *Cooperation in Agriculture* (New York, 1929), pp. 141–7. Other statistical analyses confirmed Professor Filley in the view that, "All things considered, it seems evident that many men have over-emphasized the effect of the heavy sale of wheat upon the market at the time of threshing" (*ibid.*, p. 145).

strong presumption that, taking one year with another, the holding of grain from fall till spring would yield no certain margin of profit above carrying charges.

Whatever the analysis summarized above did or did not prove, the fact is that it was not available in the early 1920's when the pools were organized. Furthermore, if it or similar studies had been available they would have made no difference. The conviction was widespread among western farmers that wheat prices were unduly depressed in the autumn months and that such a condition was inescapable under the open market system. The various royal commissions which investigated the marketing of wheat in the inter-war years accepted with little question the tacit assumption that the open market system was to be continued. Starting from that assumption they directed their attention toward establishing proof of the desirability of maintaining the freedom of futures market operations under open market or competitive price conditions. The farmers of the period, however, denied the basic assumption itself. It was their firm conviction that the competitive price system ought *not* to be continued for wheat and other grains and, starting from this assumption, a defence of futures market operations was wholly irrelevant. Royal commissions and western grain growers did not come to grips in the inter-war years because they persisted in talking about different things while apparently fully convinced that they were talking about one and the same thing.

CHAPTER ELEVEN

The Wheat Pools and the Open Market System: Origins of the Pools

THE CANADIAN WHEAT POOLS[1] were organized in the early 1920's, the Alberta wheat pool commencing operations in 1923 and the Saskatchewan and Manitoba pools in 1924. Although separately organized, each in its own province, the pools were nearly identical in form and conducted their pooling operations through a single, jointly owned and controlled Central Selling Agency which was organized in 1924. It is thus possible to speak of the organizations collectively, for the period during which they conducted pooling operations, as the Canadian wheat pool, or, simply, the pool or the pools. The pools stopped pooling wheat or other grains on a contract basis in 1931[2] and, despite the survival of the Central Selling Agency, each went its separate way, operating essentially as a co-operative elevator company. The period from 1923 to 1931 therefore forms a distinct interval in the history of Canadian agricultural co-operation. The main events in the history of that interval have been outlined in detail in a number of studies.[3] Our concern at this point is with the development of grain marketing policy in Canada in the interwar years and with the relation of this development to the general structure of Dominion policies relating to the West.

Pooling is the commingling of funds or resources or of the production and marketing of goods and services. If the product involved is solely

---

[1]Alberta Co-operative Wheat Producers Limited; Saskatchewan Co-operative Wheat Producers Limited; and Manitoba Co-operative Wheat Producers Limited. The three provincial pools as listed here were incorporated provincially. Canadian Co-operative Wheat Producers Limited, the Central Selling Agency, was incorporated by the federal government in 1924.

[2]The circumstances leading to the abandonment of contract pooling of wheat in June, 1931, are outlined briefly in chapter XIII below.

[3]See particularly H. S. Patton, *Grain Growers' Coöperation in Western Canada* (Cambridge, Mass., 1928), and by the same author, "The Canadian Wheat Pools in Prosperity and Depression" in Norman E. Himes, ed., *Economics, Sociology and the Modern World: Essays in Honor of T. N. Carver* (Cambridge, Mass., 1935); W. A. Mackintosh, *Agricultural Cooperation in Western Canada* (Kingston, 1924); H. A. Innis, ed., *The Diary of Alexander James McPhail* (Toronto, 1940), hereafter cited as *McPhail Diary; Report of the Royal Grain Inquiry Commission, 1938* (Ottawa, 1938).

that of a single producer the process creates no special problems and is taken for granted. If, however, the product of more than one business unit is thrown into a common fund or pool, the question of how to apportion the total proceeds must arise, for the monetary yield of the product supplied by the individual producer cannot be segregated from the common fund except on the basis of some wholly arbitrary rule. Nor, of course, is it possible to segregate the portion of the total expenses of the pool which are attributable to consignments from individual producers except on the basis of some equally arbitrary rule. The pooling of produce has as its most essential feature, therefore, the establishment of a set of rules for allocating the total net returns of the operation among the business units whose produce is thrown into the common fund or stock.

While, in its literal sense, the term "pool" signifies the article or economic quantity of whatever sort that may be pooled, in ordinary usage it has come to mean the agency or organization which conducts its business on a pooling basis. However, it can be recognized that, looked at from any point of view, the pools and pooling characteristic of agricultural co-operation in North America in the 1920's were by no means peculiar to the economic life of that decade or to the agricultural sector of the economy in that or any other period of time. The industrial pool of the late nineteenth century symbolizes for American economic historians one of the earliest and most impermanent of the forms of industrial combination.[4] The traffic pool of the ocean shipping line and the pool train are familiar to all.

An excellent illustration of the difficulties of defining a pool in the field of agricultural co-operation appears in S. W. Yates's study, *The Saskatchewan Wheat Pool*.[5] Mr. Yates sought information on the origins of the idea from various persons who, like himself, had been closely in touch with its development in western Canada. Mr. R. D. Colquette, long associated in an editorial capacity with the *Country Guide* and its predecessor, the *Grain Growers' Guide*, expressed the opinion that the terms "pool" and "pooling" are "not as definite and clear-cut concepts as they are popularly supposed to be."[6] In Colquette's view "the use of

[4]Describing the situation in the United States, Professor Ripley said in 1916: "The pool is probably the oldest, the most common, and at the same time the most popular mode of obviating the evils of competition. Industrial pools, in fact, appear at every stage of our economic growth since the Civil War." As cited in H. C. Filley, *Cooperation in Agriculture* (New York, 1929), p. 114.

[5]Edited by Arthur S. Morton (Saskatoon: United Farmers of Canada, Saskatchewan Section, Limited, 1947), pp. 11–12.

[6]*Ibid.*, p. 11.

the word 'pool' should first be confined to those organizations which do not buy the product of their members at the current market prices, and [which] distribute excess profits, if any, on a basis of patronage . . . in most cases paying an initial advance payment which is less than the current price. . . ."[7] On the basis of this definition Mr. Colquette was impelled to point out that pooling had had a long association with agricultural co-operation and agricultural marketing prior to its development in the 1920's. In elaboration of this point he said:

In Prince Edward Island an egg circle movement was inaugurated, in 1913, which pooled each week's returns, but did not use the term "pool." The same thing happened with fruit in British Columbia. In Ontario there are hundreds of privately owned cheese factories, the product of which is sold every week or every two weeks, and the proceeds distributed to the patrons according to the price received for the whole. This has been going on for sixty years, though such privately owned cheese factories are not co-operative in any sense of the term. In California the fruit marketing organizations from the earliest days have been pooling, and all over the United States there are marketing organizations which have been operated on the same basis for scores of years. *In fact, any organization which actually markets the products of its members or patrons in bulk, and not each individual's products separately must, perforce, use the pooling system.*[8]

The essential features of the wheat pools and of the pooling period may be summarized. First, the pools represented a producer-owned and producer-controlled alternative to the open market system for the disposal of Canadian wheat. They were the first co-operatives to aspire to this position in the Canadian grain trade and since the cessation of their pooling activities in 1931 neither they nor any other farmers' organizations have attempted to repeat the experiment. Second, in the view of the wheat producers the pools represented a second rather than a first choice of instrument for the accomplishment of desired purposes. The growers' first choice by a considerable margin was for a governmental monopoly agency which would provide not an alternative to, but an exclusive replacement for, the open market system.

A third general point that can be noted in relation to the pooling period is that at no time since the organization of the pools has the open market constituted the sole channel for the marketing of Canadian wheat as it did before 1917 and for the crop years 1921–2 and 1922–3. From 1923 to 1931 the open market survived as an alternative channel for the disposal of Canadian wheat in competition with the pools. From 1931 to 1935 the operation of the open market was greatly modified

[7]*Ibid.*
[8]*Ibid.*, pp. 11–12. Italics added.

and attenuated by the existence and activities of a governmental liquidation and stabilization agency. From 1935 to 1943 the open market
again constituted an alternative to the newly constituted Canadian
Wheat Board. Deliveries to the Board were voluntary rather than compulsory and even on that basis were permissible only under certain
restricted conditions. Since 1943 the open market channel has been
closed entirely in Canada for the marketing of Canadian wheat and the
Canadian Wheat Board has operated as the exclusive agent between
the grower and the exporter.

The above points have logic in reference to the analysis set forward
in previous chapters and merit consideration particularly with regard
to the question of governmental policy.

Earlier chapters in this volume have stressed the laissez-faire element
in the agricultural policy of the Dominion government before and during
the First World War. The replacement of the open market system for the
disposal of Canadian wheat during the war was purely a matter of urgent
expediency which involved no modification of basic Dominion policy
on the marketing of agricultural products. The summer of 1920 found
the Dominion government determined not to postpone further the re-
introduction of the "proper" type of policy and institutions for agricultural markets. In terms of wheat marketing policy this meant the
restoration of the open market centred around an autonomous Winnipeg
Grain Exchange with facilities for both cash and futures trading.

The market experience of the Canadian wheat grower from 1917 to
1920, however, first with the Board of Grain Supervisors and later with
the Canadian Wheat Board, provided a convincing set of arguments
against the restoration of the open market system. The observation of
the serious reaction on the price of wheat which coincided with the
termination of the Board's activities was sufficient to remove any linger-
ing doubt which the wheat grower might otherwise have had on the
matter. In his experience, the Board of Grain Supervisors and the Wheat
Board were the first agencies of any sort to replace the open market
system. Apparently there was nothing sacrosanct about the system and
no insurmountable barrier against its removal. Both absolute and relative wheat prices were unexpectedly favourable to western growers during
the tenure of these agencies, and on the discontinuance of the Board,
the decline in prices, both absolutely and relatively, was catastrophic.

Table VII presents data for the decade between the outbreak of the
First World War and the organization of the pools. These data make
clear the main features of the experience of western farmers with prices
during the years when the open market was replaced by government

TABLE VII

AVERAGE ANNUAL WHEAT PRICES AND INDEX NUMBERS OF
WHOLESALE PRICES IN CANADA, 1914–23

Year	Average annual price per bushel, No. 1 Northern, Fort William and Port Arthur (cash basis)	Index numbers of wholesale prices (1914 = 100)
1914	$1.00	100.0
1915	1.28	107.5
1916	1.38	128.6
1917	2.20	174.5
1918	2.22	194.5
1919	2.37	200.4
1920	2.51	238.0
1921	1.65	167.9
1922	1.23	148.5
1923	1.08	149.0

SOURCE: Dominion Bureau of Statistics.

agencies. Since the average price of No. 1 Northern Manitoba wheat at Fort William was $1.00 per bushel in the open market in 1914, the absolute data for succeeding years form in themselves a readily interpreted index. During the operation of the Board of Grain Supervisors the price of wheat was considerably more than double. The Board, in fact, fixed the price of wheat, basis No. 1 Northern, Fort William, at $2.40 per bushel in July, 1917, and reduced it to $2.21 in September of the same year. The price was held approximately at the latter level throughout the continuance of the Board of Grain Supervisors. The Wheat Board, charged with the exclusive disposal of the 1919 crop, did not fix the price of wheat but paid an initial payment on delivery by the farmer and a final payment on the liquidation of its entire holdings. The initial payment was $2.15 per bushel, basis No. 1 Northern, Fort William, and this with an interim and a final payment totalling 48 cents brought the yield for the 1919 crop to $2.63 per bushel for wheat of that grade and location. The data in the Table, being on a calendar year basis, do not coincide exactly with these crop year figures. This discrepancy is particularly important in the interpretation of events in the years 1920 and 1921, for the sharp decline in prices occurred after the middle of 1920 and coincided with the discontinuance of the Wheat Board.

Obviously, annual averages such as those portrayed in Table VII conceal the extremes of seasonal variation. Cash wheat, basis No. 1 Northern, Fort William, held at a peak price of $3.15 per bushel for the months of May, June, July, and part of August, 1920.[9] With the opening of the new crop year and the restoration of the open market, the price fell sharply and persistently to low points of $2.70¾, $2.48½, $2.17½, and $1.78½ for the months of August, September, October, and November, 1920, respectively. By December, the *average* for the month, $1.93½, was below $2.00 for the first time since early in 1917. The decline continued less sharply throughout the first six months of 1921, but with the marketing of the new crop the price again receded rapidly. During September, 1921, cash wheat in the Winnipeg market fell to $1.35⅛ per bushel and averaged but $1.48⅛ for the month. A year later, during September, 1922, cash wheat in Winnipeg fell below $1.00 per bushel for the first time since the early months of the war.[10]

Table VII indicates the existence of sharp changes in price, relative as well as absolute, affecting the wheat grower in the decade preceding the establishment of the pools. The wholesale price index of the Dominion Bureau of Statistics lagged substantially behind the price of wheat both on its rise and on its subsequent fall. For each of the years from 1915 to 1920, inclusive, the price of wheat was higher in relation to the base of 1914 than was the wholesale index. This fact is the more significant when it is noted that the wholesale index included wheat as one of its components. Therefore the lag in the prices of other commodities as compared with wheat was more pronounced than the index would suggest. Throughout the First World War and until mid-1920 the terms of trade moved increasingly in favour of the Canadian wheat grower. After that time, however, the relative change was sharply in the opposite direction. The prices of wheat and other agricultural products collapsed and the prices of commodities manufactured or processed in the industrial sector of the economy displayed a comparatively high degree of rigidity. The terms of trade moved disastrously against the agricultural producer.

It is not to be inferred from these facts that the displacement of the open market system in 1917 and its restoration in 1920 were exclusively or even primarily responsible for the drastic changes in the price of wheat in the pre-pool decade. The disturbances created by the war in

[9]This information was secured from Sanford Evans Statistical Service, *Canadian Grain Trade Year Book* (Winnipeg, annually), 1921–2, pp. 21–43. The data in the next sentence of the text represent the *low* quotation for each month and not the average.

[10]Low for the month was 96½ cents and average for the month was 99⅞ cents. The price rallied slightly and did not fall below $1.00 again until October, 1923, after which the monthly average remained below $1.00 for seven months. See *ibid.*, 1923–4, pp. 27–49.

the conditions of demand and supply relating to breadstuffs ought not to be overlooked, although the complexity of the forces involved renders the measurement of the relative significance of these various disturbances extremely difficult. In terms of wheat marketing policy, the important consideration is the interpretation which the growers placed upon the observable facts at the time and throughout the immediately succeeding years. Contemporary sources make it clear that the growers regarded the removal and eventual restoration of the open market system as of very considerable significance in the cycle of prices and purchasing power through which they had passed in the decade which ended in the early 1920's. At a meeting of the Saskatchewan Grain Growers' Association in October, 1920—at which time the price of wheat had fallen in an interval of two months from a peak price of $3.15 per bushel to $2.17½—H. W. Wood insisted that the price could have been maintained at $2.50 per bushel at least, had the Wheat Board remained in operation.[11] In articles in the *U.F.A.*, the official organ of the United Farmers of Alberta, Mr. Wood argued that the pre-war basis of the value of wheat had been its feed value and that after the war, with no basis available and in the absence of efficient marketing, wheat had gone down in price to the level of coarse grains.[12]

For three years after the Canadian Wheat Board was disbanded in 1920 there was great uncertainty in the West concerning the form of wheat marketing organization which should be provided for the post-war years. The one point on which there was full agreement among wheat growers was that the open market system ought not to be restored to its pre-war position as the exclusive channel for the disposal of the Canadian crop. The obvious possibility in view of the recent past was that the Dominion government might be persuaded to reconstitute the Wheat Board and to continue its monopoly. In the hope that this possibility might be realized, the western Grain Growers' associations, the national Canadian Council of Agriculture, and the western legislatures made repeated representations to the federal government and did not abandon their efforts in that direction until 1923, when it was finally made abundantly clear that the Dominion government was not to be persuaded.

Meanwhile, as early as the autumn of 1920, the farmers' associations had given consideration to a second possibility, that of organizing some

[11]Patton, *Grain Growers' Coöperation in Western Canada*, p. 199. H. W. Wood had been a member of the Canadian Wheat Board and was currently president of the Canadian Council of Agriculture.
[12]*McPhail Diary*, p. 45.

form of commodity pool controlled by the growers. In December, 1920, the Wheat Markets Committee, which had been established by the Canadian Council of Agriculture to consider and report upon the matter of a marketing organization, proposed the establishment of a five-year contract pool, to be known as the United Farmers' Grain Corporation, without elevators, the existing farmers' elevator companies to handle grain for the Corporation as they had for the Canadian Wheat Board. On further consideration this proposal appeared to be impracticable and was abandoned in favour of a renewal of pressure on the Dominion government for the restoration of the Wheat Board.

The reference to the farmers' elevator companies raises the question of the extent to which western grain growers were able to rely on these companies to solve the wheat marketing problems of the inter-war years. By the end of the First World War the two farmer-owned elevator companies—the United Grain Growers and the Saskatchewan Co-operative Elevator Company—financially strong and well integrated, handled approximately one-quarter of western grain and provided competitive facilities at approximately 45 per cent of the local shipping points in the West.[13] The contribution which they had made toward the removal of abuses could not be denied. Nevertheless they proved unwilling or unable to adapt themselves to the insistent demands voiced by the growers in the 1920's for fundamental reform in marketing. This fact was not recognized immediately after the war but was not long in becoming apparent. The proposal advanced by the Canadian Council of Agriculture in 1920 identified these companies as convenient auxiliary instruments rather than standard-bearers of reform.

Even before the end of the First World War, western grain growers were critical of the co-operative elevator companies, and their criticisms gathered force and cogency in the post-war years. Professor Mackintosh has summed up their complaints as follows:

It has been assumed that they [the co-operative elevator companies] are cooperative. Yet no charge is more frequently brought against the companies than that they are simply money-making corporations masquerading under the cloak of cooperation. . . . Others who do not condemn the companies as fakes yet protest that they are not different from other private companies, that they have made huge profits, that they follow the same practices, that they are members of the Grain Exchange and make use of the speculative market. . . .[14]

[13]The farmers' elevator companies handled one-third of all western grain in the crop year 1915–16; in the early inter-war years they were handling from one-fifth to one-quarter of western inspections. Patton, *Grain Growers' Coöperation in Western Canada*, pp. 325, 189.
[14]*Agricultural Cooperation in Western Canada*, p. 90.

As proof of the charge that the co-operative elevator companies were not really co-operative it was pointed out that they did not pay patronage dividends but instead paid substantial dividends on share capital and accumulated strong financial reserves. The real crux of the matter in the post-war years, however, was that the co-operative elevator companies had identified themselves with the open market system of marketing, had adapted their operations and their policies to meet its demands, and could therefore not be expected to work actively for its abolition. By the early 1920's the abolition of the open market system had come to be regarded by a substantial proportion of western grain growers as indispensable.

The organization of the wheat pools marked the commencement of a new phase in agricultural co-operation in western Canada. The pools were conceived as a new form of co-operative organization with radically new methods of operation, and their goal in terms of modifying the price system was appreciably more ambitious than the objectives of previously organized Canadian agricultural co-operatives. It must be made clear, however, that in terms of basic function the pools were designed to perpetuate the tradition which had been fully established in the early years of the present century, the tradition that co-operative organization might serve as an effective expression of the farmers' dissatisfaction with their place in the price system. The objective of the pools differed in degree rather than in kind from those of earlier co-operative organizations.

A detail of some interest is that the co-operative grain marketing companies which were organized before 1914 were concerned with reducing the grower's vulnerability not so much in the markets in which he sold his produce but, rather, in the markets in which he *bought* the services of middlemen.[15] The pools, on the other hand, directed their attention primarily to the markets in which the farmer *sold* his produce. They by no means ignored the farmer's purchasing markets, for they provided a full set of facilities, including country and terminal elevators, from the local shipping point to the domestic and overseas milling centres where Canadian wheat is consumed. The provision of these facilities, however, was only partly in the interest of economy of operation and was mainly designed to assist the pools in controlling prices. In short, the pools were not organized to improve upon the

[15]In markets which, looked at from the farm-enterprise point of view, would be described in present-day theoretical analysis as "input" markets. The pools, concerned on the other hand with the markets in which grain growers sold their produce, were correspondingly interested in farmers' "output" markets.

storing and export services which were already offered by the farmers' elevator companies. It is doubtful indeed whether the pools *could* have improved upon these facilities to any appreciable extent.

In reference to selling markets, the wheat grower is interested in the net price to be obtained at local shipping points. Those who advocated a fundamental change in the marketing structure in the early inter-war years assumed without question, as had the co-operative organizers of previous decades, the desirability of maximizing the local net return per bushel. But there the identity of assumptions ended. Beyond and above the local price for wheat is the "spot" or terminal price, and beyond that is the price in overseas markets, the latter ordinarily symbolized by the "Liverpool price" for wheat. The local price at any geographic point at any one time depends on the spreads existing between the various levels in the price structure and on the price at Liverpool as well. Individuals and organized groups whose vital concern is uniformity with the prices at local shipping points in the wheat producing area may therefore have widely differing interpretations of the causes of unsatisfactory local prices and at least as wide a variety of opinions on how the unsatisfactory condition may be corrected.

It is a simple truism that if the price secured by the Canadian grower for his wheat is unduly low, this circumstance must arise from an excessive spread between local and overseas prices, or from an unsatisfactory overseas price, or from a combination of both.[16] This truism is helpful toward an understanding of the price policies of grain growers' co-operatives. It is clear that a co-operative organization might centre its attention either on the spread in prices between local and overseas markets (whether attributable to inefficiency or monopoly), or on the price paid by the overseas importer or miller. It is also clear that in directing attention to price spreads there may be considerable difference of opinion as to what spreads are significant and capable of being appreciably reduced. From such differences of opinion a diversity of co-operative purpose, activity, and even form may arise.

This analysis may appear to point to a simple, clear-cut distinction between the elevator companies and the pools, the former centring their attack on price spreads and the latter directing their attention toward alteration of overseas prices. The distinction cannot be so sharply drawn. The policies of the elevator companies were not exclusively concerned with prices, but primarily with the provision of adequate elevator facilities at local and terminal points. The price features of their policies were prominent, nevertheless, and were exclusively concerned with the

[16]See chap. VII, p. 104.

possibility of narrowing the handling margin between local and export points.

The pools, on the other hand, were concerned almost exclusively with price policies and, as long as their method of handling grain continued, devoted their energies to providing elevator facilities only to the extent necessary to render their price policies more effective. In the price policies of the pools, in contrast to those of the elevator companies, there was a pronounced concern over the final selling price for Canadian wheat in overseas markets. The Canadian grower had seen these prices rise within the span of a few years to more than treble their pre-war level and fall again even more quickly to approximately the 1914 figure. He could not but be impressed by this demonstration of unexpected flexibility and, at the same time, he could not but observe that a continued whittling down of handling margins offered little monetary gain in comparison with that which might accompany a substantial increase in the world price of wheat. Those who advocated a continuation of the Wheat Board's method of marketing and who eventually turned to organize and support the pools as a viable alternative were aiming at bigger stakes than the further reduction of handling margins. The possibility of influencing world wheat prices in ways which would prove financially beneficial to the growers of wheat could never be entirely out of the minds of those who advocated a wheat board or a wheat pool.

We cannot infer from this, however, that the pools' price policy was primarily directed toward the control of world wheat prices. One of their significant experiences was a deeply seated conflict over price policy in the ranks of their leaders. There were those who persistently felt and argued that the great opportunity of the new type of co-operative organization exemplified by the pools lay in the exertion of effective pressure on prices in the ultimate selling markets. There were others who were not convinced that such a possibility lay within the power of a voluntary farmers' marketing organization and who would cast the pools in a more modest role. During the early years of the pools the counsel of the latter group prevailed. During the later years, the years of crisis which ended with the abandonment of the pooling method, the counsel of the former group was advanced with increasing vigour and won increasing support in the ranks of the membership against the opposition of leaders who held stubbornly to more moderate views. The issue was eventually joined on the question of compulsory *versus* voluntary pooling. The progress and outcome of this struggle is outlined in chapter XII.

The group whose counsels were predominant in the early years had at least one thing in common with the organizers and supporters of the co-operative elevator companies: they believed that the great opportunity for co-operative service to the producer lay in narrowing the spread between local and overseas prices. The elements of that spread which they sought to narrow were, however, quite different. The elevator companies had attacked handling margins and the spread between street and terminal prices, the assumption being that both these elements were unduly inflated by monopoly. The pools concerned themselves very little with either handling margins or inter-regional price spreads, assuming that a persistent striving after efficiency of operation would yield the maximum gains that could be anticipated in these areas of endeavour. Their primary attention and efforts were directed toward the narrowing of *seasonal* rather than *geographic* spreads. In the pointed terminology of the day these spreads were seldom described as "seasonal" but almost unfailingly as "speculative." They were not attributed to the seasons nor to the monopolistic position of line elevator companies but were regarded as wholly due to the speculator and to speculative activity. The futures market and with it the whole open market system apparently provided the mechanism within which the speculator created and took advantage of seasonal spreads at the expense of the "producer" of grain.[17]

Early in 1921 the Premier of Saskatchewan addressed a number of questions to James Stewart and F. W. Riddell to secure their advice on ways of dealing with the wheat marketing problems of the day.[18] Mr. Stewart had been president of the Wheat Export Company, a member of the Board of Grain Supervisors, and chairman of the Canadian Wheat Board. The Premier's first question was whether it was "possible for any kind of pool comprising less than the whole of the western wheat crop to market the crop to the same advantage from the producers' point of view as a system of national marketing of the whole crop by a Canadian Wheat Board."[19] Messrs. Stewart and Riddell answered in the negative, that is, they were of the opinion that a partial pool would of necessity be less advantageous to the producer than would a monopolistic pool or wheat board. As a second-best alternative to the all-inclusive pool or board they favoured a completely voluntary pool, and for this preference they outlined their reasons at considerable length.[20]

[17]See chap. x, pp. 186–95.
[18]For questions and answers referred to at this point see James Stewart and F. W. Riddell, *Report to the Government of Saskatchewan on Wheat Marketing* (Regina: King's Printer, 1921).
[19]*Ibid.*, p. 3.
[20]*Ibid., passim.*

The second question was: "What advantages to the producer over the present [open market] system would there be in any kind of pool which comprises only a portion of the crop?"[21] The reply given by Stewart and Riddell summarizes concisely the rationale of the wheat pool movement of succeeding years and is therefore most pertinent to an analysis of the price policy of the pools. It is not so much that their statement was publicized or widely quoted, but that it embodied what growers throughout the West had come to regard as a group of obvious and significant facts. Under the heading, "Advantages of Pooling Systems," Messrs. Stewart and Riddell replied in part as follows:

Other things being equal, pooling systems have the following advantages over the present system of marketing:
(1) Statistics show that under the present system seventy to seventy-five per cent. of the crop is thrown on the market during a period of three months. Under a pool, with proper financial and other support, the movement of the crop would be more evenly spread over the whole crop year, thereby undoubtedly avoiding gluts of grain, and consequent depression of prices, which usually occur during the first three months of the crop movement.
(2) The more evenly regulated movement would stabilize prices to the consumer, as well as in favour of the producer. . . .
(3) The price of flour to consumers bears a direct relationship to the prevailing price of wheat. . . . Three-quarters of the farmers' wheat is marketed during a period of three months, at the beginning of the season, when the price of wheat usually is depressed. During the remaining nine months the tendency has always been for the price of wheat to ascend to higher levels, with the result that the producer only gets the benefit of the prevailing higher prices for the remaining quarter of his season's crop. . . . [Therefore] it would be to the advantage of the consumer, as well as the producer, if the delivery of that wheat crop could be more evenly spread over the entire twelve months. If this could be done, fluctuations in prices would be lessened, and it is a well recognized fact that fluctuations in prices are detrimental either to the producer or consumer.[22]

This statement outlines the concept of the autumn glut of Canadian wheat and explains its significance for prices in what would appear to be indisputable and completely unequivocal terms. Western growers had no alternative but to market the bulk of their crop immediately after threshing: they needed cash, the storage space on their farms was inadequate, and the winter months were severe. The "dumping" of the Canadian crop, forcibly summarized in the statistical expression that three-quarters of the annual product was thrown on the market within

[21]*Ibid.*, p. 3.
[22]*Ibid.*, pp. 15–16.

one-quarter of the crop year, led inevitably—or so it appeared—to unduly depressed prices throughout the critical months when the grower sold the most of his output. As an alternative to the dumping of the wheat crop, which was unavoidable under the open market system, the advocates of pooling were able to propose the attractive possibility of "orderly marketing," or, as Davisson described it, "the gospel of orderly marketing as preached by Sapiro."[23] Orderly marketing, or merchandising, was to replace "the flooding of unorganized grain on the market, without the least concern as to market needs."[24]

A further point of significance in the statement made by Stewart and Riddell is that they related the price possibilities of pooling to the narrowing of the spread between producers and consumers and not at all to any possibility of increasing prices to consumers. Since "the price of flour to consumers bears a direct relationship to the prevailing price of wheat," consumers tend to secure the advantage of cheap flour for three months and are forced to pay high prices for nine months of the year. But the consumption of flour is not seasonal, and high prices are paid for three-quarters of the year for approximately three-quarters of the annual consumption of that basic commodity. Thus the argument implicit in the statements advanced by Stewart and Riddell is that consumers pay high prices for the product of three-quarters of the Canadian crop while the producers secure correspondingly high prices for only one-quarter of the same crop. The discrepancy could only be accounted for by an excessive spread which Stewart and Riddell did not designate but which the western producer identified readily as a speculative margin attributable to the workings of the futures market specifically and of the open market system generally.

The question of price policy will be dealt with at greater length in later sections of this study.[25] In terms of historical sequence the full implications cannot be said to have become clear until after the pools had been in operation for at least a year or two and until pronounced differences of opinion among the leaders brought the issue into sharp focus. It can be stated at this point, however, subject to later elaboration, that, in the years immediately preceding and following the formation of the pools, their price policy was based on the assumption that prices to consumers would, on the average throughout the year, remain unchanged but that the pools could divert to the grower a greater

23Walter P. Davisson, *Pooling Wheat in Canada* (Ottawa, 1927), p. 21.
24*Ibid.*, p. 51. "Stop dumping and start merchandising," said Sapiro in Saskatoon, Aug. 7, 1923. *McPhail Diary*, p. 48.
25See particularly chap. XII.

proportion of the average price than he could secure in the open market. This diversion, to be made possible by narrowing the over-all spread between producers and consumers, was to be secured primarily at the expense of the speculators, those persons who presumably acquired cheap wheat on the depressed markets which were typical of the fall of the year and who disposed of the wheat at the higher prices which obtained throughout the remainder of the year.

In his Preface to *The Diary of Alexander James McPhail* Professor Innis tells of his first meeting with McPhail, a meeting which took place in Regina on August 10, 1923. Sapiro was in the midst of the whirlwind campaign of evangelical oratory by means of which he transmitted such vitality to the wheat pool movement. Professor Innis spent the evening in the office of the Saskatchewan Grain Growers' Association in conversation with McPhail, Robertson, and Edwards.[26] "The discussions of that evening," says Professor Innis, "ranged over such topics as the relations between the East and West, the secession movement, and the Hudson Bay Railway, but continually returned to the possibilities of the pooling system. It was generally agreed that the pool could not raise prices but that it could introduce various economies in marketing and increase and steady the returns to the farmer."[27] This statement may be regarded as a fair summary of the early philosophy of the pool's leaders and of the movement generally. It must be recognized, however, that the economies and the increase in growers' returns which are referred to in the statement rested on possibilities of a more ambitious nature than those relating to the further paring of the elevator companies' handling margins and envisaged the much more substantial opportunity which appeared to exist for the diversion to the grower of exorbitant speculative spreads.

Although organized by growers well experienced in co-operation, the wheat pools had aims and methods unlike those of the earlier co-operatives. The Canadian Wheat Board of the 1919–20 crop year, and

[26]A. J. McPhail, G. W. Robertson, and G. F. Edwards were at that time all prominent in the leadership of the Saskatchewan Grain Growers' Association and were all active in the wheat-pool organization-campaign of 1923–4. Edwards was vice-president and McPhail was general secretary of the S.G.G.A. All were members of the organizing committee and of the provisional and first permanent boards of directors of Saskatchewan Co-operative Wheat Producers Limited. McPhail was vice-president of the Board until July 25, 1924, when he was elected president, the position which he held until his death on September 28, 1931. He was also selected president of the Central Selling Agency on its formation in August, 1924. G. W. Robertson has been secretary of the Saskatchewan Wheat Pool from its origin until the present time.

[27]Page vii.

certain developments in the field of American agricultural co-operation, provided complementary and corroborative examples which far outweighed all other influences in the determination of their structure and objectives. The Canadian Wheat Board provided an example for methods of operation and results rather than forms of organization or objectives. As a governmental agency designed to serve as the exclusive merchandising contact between the Canadian wheat grower and the Allied buying organization the Board could not, of course, be duplicated in any literal sense of the term by a farmers' organization forced to rely on persuasion rather than compulsion and required to accept the re-establishment of the open market system in overseas buying markets. A number of the Board's operational practices could, nevertheless, be duplicated or approximated by a producers' agency.

Most important was the pooling of returns on a grade basis. All growers who delivered a particular grade of wheat to the Wheat Board received the same return per bushel at Winnipeg for grain delivered to the head of the Lakes, regardless of the dates within the crop year on which their respective deliveries took place. Freight charges from the local delivery points to the head of the Lakes were not pooled but were deducted in their respective specific amounts from the first payment made to the individual grower. Once the grain was delivered to the lakehead, however, the total proceeds from the Board's sales for the entire year, less operating costs, were apportioned equally to every bushel of grain handled by the Board with the exception of fixed but narrow spreads between the various grades. The net return per bushel to individual growers varied, therefore, only because of differences in the grade of their product and in the freight charges. This simplification of the pricing procedure marked a revolutionary change from the open market system where in addition to variations due to grade and freight the grower faced the substantial uncertainty as to whether the Winnipeg price at the time of his sale would approximate the annual average or would in fact be above or below that figure by an appreciable margin. In view of the widespread conviction that wheat prices in Winnipeg in the fall of the year were far below those which obtained throughout the remainder of the year, when overseas buyers secured the bulk of their requirements and when western growers had little left to sell, the significance of the pooling of returns for the year's sales can be readily understood.

A second important and distinctive feature of the Board's operations was the centralization of control over the disposal of the entire Canadian crop in a single agency. This was the obvious and perhaps inevitable counterpart of the centralization of overseas buying which persisted after the war. With the restoration of the open market system in the purchasing markets of the world, the question of centralization or of individual sales initiative in the Canadian market was of much more significance. Any possibility presumed to exist in "orderly marketing" —a term not used to any extent in regard to the Wheat Board's activities but of marked importance in the rationalization of wheat pool policy— could only exist for an agency with an unimpaired right of disposal over whatever grain it handled. This right the farmer-owned elevator companies obviously had for grain which they bought, but they also handled much grain on storage and commission terms with the individual grower retaining title and control over its further disposition. The wheat pools assumed full control over all grain delivered to them on contract, immediately on delivery.

Finally, the Wheat Board's activities represented a venture in direct selling and, conversely, an object-lesson in the elimination of middlemen. Two agencies, the Canadian Wheat Board and the Wheat Export Company, were adequate to replace a multiplicity of interests which, under the private trade, had handled Canadian wheat from the grower to the overseas miller. The physical task and the essential costs of moving the grain from producer to consumer obviously remained, but direct selling appeared to offer the prospect of eliminating all speculative handling margins. If any doubt existed on this score the fact was evident that the Wheat Board's operations could be, and were, conducted with apparent success without the mechanism of the futures market and the "speculative system." This fact served as no other could have done to point the lesson for the western grower.

Finally, of less but by no means negligible importance by way of example, was the Wheat Board's system of payment. The grower had been accustomed to receive full payment for his grain when he sold it. The Wheat Board made an initial payment to the farmer when he delivered his wheat and issued participation certificates which entitled him to a *pro rata* share of the additional proceeds, if any, of the total pooled crop. Partly because of the novelty of this procedure and partly, it may be, due to the magnitude of the initial payment ($2.15 per bushel, basis No. 1 Northern, Fort William), there was widespread scepticism among growers concerning the value of the participation certificates. Since they were negotiable, the certificates were bought and sold on a purely speculative basis, and many farmers disposed of theirs for ready cash at prices which proved to be ridiculously low in relation to their eventual redemption value.[28] The wheat pools adopted a system of payment similar to that of the Wheat Board but with two interim payments interposed between the initial and final payments and spaced with due regard for the seasonal cash requirements of their members. Supporters of the pools may well have accepted this method of payment the more readily because of their recollection of a surprisingly favourable experience with the Wheat Board and its cash returns.

Around 1920 there was a tremendous expansion of organized agricultural co-operation in the United States. Approximately 1,800 associations were formed in that year alone.[29] The number of associations involved is of no particular significance for Canadian developments, but certain of the structural changes were of the utmost importance for the design of agricultural co-operative associations in the post-war years. The Canadian pools, not only in the grain trade but also in the marketing of livestock and dairy and poultry products, were modelled in practically all important details and certainly in all of their distinctive features on American examples. The American pool movement in its organizational stages preceded the Canadian movement by the very few years which were sufficient to impress Canadian farm leaders with the long-run possibilities of the pooling system, but not sufficient to demonstrate its more serious limitations. The agricultural marketing pools of the new type which had acquired such a vogue in the United States and attracted the favourable attention of western Canadian farm leaders after 1920 were regarded as the embodiment of true co-operation. The pooling of farm produce was an essential part of the operating policies of these associations but other characteristics were observed to be at least equally indispensable to the constitution of what appeared to be a revolutionary new type of marketing association.

The American associations were large-scale, centralized, and farmer-owned organizations, specializing in single commodities; they were organized on a non-stock, non-profit system of finance and ownership; they relied for patronage on rigid delivery contracts, voluntarily entered into by the grower but binding for a period of years, commonly five or

[28]In order to stop the unnecessary sacrifice of participation certificates by western farmers the Wheat Board made an announcement on May 5, 1920, that these certificates would be worth not less than 40 cents per bushel. They were eventually liquidated at 48 cents. See *Report of the Canadian Wheat Board*, season 1920, p. 11.

[29]R. H. Elsworth, *Statistics of Farmers' Cooperative Business Organization* (Washington: Farm Credit Administration, 1936), pp. 8–10, as cited in C. F. Phillips, *Marketing* (Boston, 1938), p. 119. See also R. B. Forrester, *Report upon Large Scale Co-operative Marketing in the United States of America* (London: Ministry of Agriculture and Fisheries, 1925).

seven; and, finally, as noted above, they marketed farm produce on a pooling basis with total payments made up of initial, interim, and final disbursements for the entire pool period.

That these characteristics when combined in individual agricultural co-operatives created an instrument with unique possibilities for the solution of agricultural marketing difficulties has been much more in doubt since 1930 than it was in 1920. Farmer-owned elevator companies by 1920 were obviously large-scale and centralized and were specialized on a commodity basis. At that point, however, their similarity with the new American model ended.

Opinions differ over which of the features of the newer type of co-operative was the most novel in terms of practical requirements. A careful student of American co-operative devolopment, Professor Filley, considered the use of the membership or patronage contract to be the most significant departure from previous practice. Commenting on this point in the late 1920's he said:

These membership contracts have been the outstanding feature of many of the more recent California co-operative associations which have been successful and have been adopted by the Burley Tobacco Growers' Association, by . . . [a number of the cotton associations], by many associations of milk producers, by associations of wheat growers, and in fact by a very large majority of the large marketing associations formed in recent years. The membership contract may be said to be the only really important addition made to the principles of the original Rochdale Pioneers in the many years that their plan of co-operation has been in operation.[30]

The fact that the newer associations were almost universally called "pools" would suggest that the pooling of produce was the feature which most clearly distinguished them from other types of agricultural associations. The newer associations *did* pool the produce which they marketed for their members, but all the available evidence indicates that pooling was but a means to an overriding end and not in itself either a sufficient means or an end. Mr. Yates divides the co-operatives of the 1920's into the federated type and centralized type, and adds, ". . . the latter . . . spread all over North America as a result of Sapiro's organizing genius and has had applied to it the handy word 'Pool'."[31] There is a significant implication in this statement: the name "pool" as applied to any of the agricultural co-operatives in the inter-war years may be regarded more as a matter of convenience than as an attempt to single out the one characteristic of pre-eminent importance.

[30]*Cooperation in Agriculture*, pp. 95–6.
[31]*The Saskatchewan Wheat Pool*, p. 11.

A letter written in May, 1923, by A. J. McPhail, secretary of the Saskatchewan Grain Growers' Association, to Mrs. McNaughton, a member of the executive, expresses his concern over the delicate position of those responsible for the policies of the Association in the matter of the marketing of wheat.[32] The farmers' associations of the three western provinces in their 1923 convention had "reiterated their demand for a Wheat Board just as strongly as in former years," and the federal government had passed legislation which would permit the formation of a wheat board by joint action of the provincial governments. However, a bill to provide for participation by the Manitoba government had been defeated by a narrow margin in the Manitoba Legislature within the week before McPhail wrote. The likelihood that Saskatchewan and Alberta would or could go ahead with the formation of a wheat board for the marketing of the 1923 crop was slight. Nevertheless, so long as there was any possibility that they would, McPhail felt that it was questionable "whether any officer of the Association should, at this time, come out flat-footed for the creation of a co-operative marketing organization . . . [since] the people might take the stand . . . that it was a move to kill the Wheat Board." On the other hand McPhail was firmly of the opinion that the grain farmers' marketing problems could be solved only by organizing a marketing agency owned and controlled by themselves, and that if the associations did not come out firmly for such an agency some one else would, "and as a result our own organization can be accused of having been asleep at the switch."

Other paragraphs in McPhail's letter show clearly the extent to which western farm leaders kept American co-operative developments under observation in the early 1920's, and also indicate the features of the new American co-operatives which appeared to be relevant to Canadian needs. McPhail comments on an interview with certain co-operative groups in the United States:

When we visited the offices of the North-West Wheat Growers' Association[33] [in Chicago], the men whom we saw there seemed to be very optimistic regarding the future of the co-operative plan, and were of the opinion that

[32]*McPhail Diary*, pp. 39–40.
[33]The Northwest Wheat Growers, Associated, was organized as a state pool in the state of Washington in 1920 and the idea spread to other states in the northwest. These pools had membership contracts of seven years' duration. During 1924, however, the various state pools in the group were voluntarily disbanded. See Filley, *Cooperation in Agriculture*, pp. 127–8. Patton (*Grain Growers' Coöperation in Western Canada*, pp. 210–11) suggests an important point of contact between Canadian and American farm leadership in the early inter-war years. "The [*Grain Growers'*] *Guide*," he says, "sent a staff representative to investigate farmers' coöperative marketing organizations in California and elsewhere, and

before a great while the bulk of the wheat grown in the United States would be marketed through a centralized co-operative farmers' organization. They are also looking forward to the not distant future when the Canadian wheat will also be marketed through a similar agency. When that time arrives there will be little difficulty, with organizations complete in the two countries, in having them co-operate in the control of the merchandising of this product.

McPhail was impressed by the optimism of certain American farm groups concerning the future of "the co-operative plan" and expresses his belief that the plan must be adopted by Canadian wheat growers, by creating a "centralized co-operative farmers' organization" for the marketing of their wheat. Two points are worthy of comment. First, there is no suggestion in his letter that the two large farmer-owned elevator companies which existed in Western Canada in the 1920's offered any prospect for the implementation of "the co-operative plan." Although they were centralized, genuinely farmer-owned, and at least nominally co-operative, the co-operative plan as envisaged by McPhail required a "centralized co-operative farmers' organization" with something in addition, the ability to work toward "the control of the merchandising of this product [wheat]." In the second place, McPhail's comments contain no reference to a pool or to the pooling process as part of the co-operative plan. The American associations to which McPhail referred marketed produce on a pooled basis, but this policy along with others was apparently regarded as merely incidental to the overriding purposes of these associations, the co-operative merchandising of the members' produce.

Along with the forms of organization and operating policies which the Canadian pooling movement derived from the American example, it imported much of the rationalization and terminology of its American counterpart. The expression "orderly marketing," by no means free from ambiguity and, in fact, subject to serious misconception, was thus adopted in Canada as a phrase which purported to summarize and explain the chief objectives of the agricultural pools. This expression requires further elaboration and this will be provided in chapter XII. At this point, orderly marketing as it applied to the co-operative movement may be defined in the broadest sense as the control of the movement of produce to market by a collective agency. This definition leaves

---

during 1920 and 1921 published an extended series of articles designed to inform its readers regarding the methods and achievements of non-profit commodity pool organizations—a form of coöperation with which Western grain growers were as a whole unfamiliar. Owing, however, to the insistent demand of the great majority of farmers for immediate concentration upon the restoration of the Wheat Board, the coöperative plan remained in abeyance until the latter part of 1923."

undefined all the controversial issues which inevitably arise in any study of the interpretation and implementation of an orderly marketing policy. It may, however, serve to indicate the individual and collective necessity for incorporating the features of American co-operatives, known as pools, into the Canadian wheat pools.

In what way did the various features of the pool type of organization combine to promise a degree of collective control over the movement of merchandise to market? It can readily be shown that the features of most significance in this regard were large-scale centralization, pooling, and the use of the long-term patronage contract. The necessity for centralized organization on a large scale was, in the early 1920's, greater in the United States than in Canada. Canadian grain marketing co-operatives were already large and centralized and there was no suggestion that small organizations or associations with decentralized control could cope with Canadian grain marketing difficulties. In the United States, however, there were two kinds of large-scale marketing co-operatives in the agricultural field, the federated and the centralized, and despite the many evident advantages of the former type, it obviously could not exercise the degree of marketing control desired and anticipated by Canadian growers in the post-war years. In the California Fruit Growers' Exchange, the outstanding example of the federated type, and one with many recognized successes to its credit, the local associations retained complete autonomy in the determination of the time and place of the disposal of members' produce; there were approximately two hundred of these local associations, and it was evident that the central exchange of the federation could not exercise authority over them in the movement of the produce through marketing channels. Observation of these limitations predisposed Canadian farmers strongly in favour of keeping their organizations centralized and giving the central agency complete discretion in the marketing of members' produce.

The pooling method, in addition to possessing obvious advantages associated with the equalization of returns among members of the organization, was almost indispensable to the control of marketing movements by a centralized co-operative organization. Under the pooling arrangement the centralized marketing association accepted the produce of its members under a general commitment to market the whole to the best advantage, but at *its* discretion rather than at the discretion of the individual grower or of any local association. Immediately upon delivery of produce, then, the pooling organization assumed full responsibility for its further disposal and might direct it forward to any market at any time, or retain it if retention appeared to be the preferable action. The membership contract bound the member to deliver his produce exclusively to the co-operative association and thus assured the association of a volume of business—variable with crop conditions, of course— regardless of competitive attractions of alternative marketing agencies.

It is evident that a variety of influences of widely diverse origins moulded the western Canadian wheat pools. No attempt has been made here to trace the sequence of the organizational process or to comment upon the significance of personalities and individual events. These have been amply outlined in other studies. Our concern is rather with the formulation and development of policy, to which we turn more particularly in the following chapter with the conviction that an outline of formative influences is prerequisite to an appreciation of policy and its objective.

CHAPTER TWELVE

Marketing Policies
of the Canadian Wheat Pools

THE PREAMBLE to the growers' contract with the individual provincial wheat pools read in part as follows: ". . . the undersigned Grower desires to co-operate with others concerned in the production of wheat . . . for the purpose of promoting, fostering and encouraging the business of growing and marketing wheat co-operatively and for eliminating speculation in wheat and for stabilizing the wheat market. . . ."[1] The preamble to the agreement signed by the provincial pools for the creation of the Central Selling Agency repeated these purposes specifically. It is clear that the pools proposed to foster co-operative wheat marketing, to eliminate speculation in wheat, and to stabilize the wheat market. While some approach to an understanding of the pools' marketing policies could be made by an elaboration of these concepts, this does not appear to be the best way of presenting an analytical synthesis of their activities.

Pool marketing policy may more profitably be considered as concerned primarily with the pursuit of direct and orderly marketing. This statement offers no clue to the many elements implicit in these terms but nevertheless provides a basis for further analysis. The pool type of cooperative structure—large-scale and centralized—and the pool method of handling grain, were prerequisite to both direct and orderly marketing as envisaged by the leadership of the Canadian wheat pools. Only when a single large organization possessed unqualified discretion in the disposal of the crops of many thousands of individual producers would it be worth while to organize and maintain the elaborate structure of overseas agencies necessary for the establishment of direct contact with the milling consumers. Only under the same circumstances could orderly marketing, in any sense of the term, be attempted. The contract signed by the growers gave to the provincial association exclusive agency rights in the growers' disposable wheat crop with the right to "sell the said wheat to millers, brokers or others, within or without the province, at

[1]From text of growers' contract, Alberta Co-operative Wheat Producers, Limited, cited in H. S. Patton, *Grain Growers' Coöperation in Western Canada* (Cambridge, Mass., 1928), pp. 429–34.

such time and upon such conditions and terms as it may deem fair and advisable."[2] The provincial pools in turn created the Central Selling Agency as exclusive agent for the administration of the powers held by the provincial organizations under contract with their individual members.

Pooling might, of course, be justified regardless of its indispensability to direct and orderly marketing. The pool method of payment removes inequality of monetary returns per unit of product as between individual members of the association. All those who deliver a particular grade of product within the pooling period (the crop year in the case of wheat), secure the average net price realized by the organization for that grade during the pooling interval. The system of spaced or deferred payments, it may further be argued, contributes toward greater stability in rural finance and serves to reduce the grower's reliance on the banks for working capital. These results, significant as they may have been, however, were regarded by western wheat growers as of secondary importance and were insufficient to justify continuance of the pool procedure after its major objectives had proved impracticable.

Direct marketing might be considered as merely one element in the pools' programme of orderly marketing. Whether or not this is the best way to regard it, there is little difficulty in explaining and rationalizing the activities of the pools in pursuance of their intention to market directly. Their intention was essentially to integrate forward from the producer, to avoid dealing with terminal buyers, lake shippers, exporters and foreign importers, and to establish domestic and overseas agencies which would deal directly with the millers of Canadian wheat.

The main objectives of such a procedure are obvious. The elimination of the middleman could readily be interpreted as leading inevitably to the saving of the middleman's profit, if not to a reduction in the actual costs of performing his necessary functions. Under direct selling, the farm price might therefore be expected to be higher than under open market conditions simply because of greater economy of operation and consequent reduction in handling margins. More particularly, however, the middleman group who handled wheat under open market conditions were thought to include a substantial speculative element, and to their legitimate handling margins they reputedly added a substantial if indeterminate speculative margin. This margin would be non-existent under pooling operations, and farm prices would be the equivalent of world prices less only an irreducible minimum of merchandising costs. Direct selling, therefore, was essential to the important objective of the

[2]Section 16, cited in *ibid.*, p. 432.

elimination of the speculator. Finally, direct selling assured close and continuous contact between the farmers' agency and the milling and consuming centres of the world. Market conditions might thus be accurately observed, reported, and interpreted. The remote observation and haphazard market action of poorly informed agricultural producers operating individually would give way, so it was argued, to skilful merchandising procedures similar to those practised with such obvious success by the great industrialists of the urban communities.

The pools, in pursuance of their direct marketing objective, established an elaborate but well integrated and comparatively economical structure of agencies. The growers were organized for administrative purposes on a provincial basis, and the Central Selling Agency was established in Winnipeg as the exclusive marketing organization for the provincial associations. It maintained a western sales agency at Calgary for the disposal of the Alberta product. It opened offices at Fort William, Port Arthur, Vancouver, Toronto, Montreal, and New York. Within the first few years of its existence it had created twenty-eight agencies in fifteen importing countries including particularly the British Isles and France, but also Germany, Holland, Belgium, Sweden, Denmark, Switzerland, Italy, Greece, Portugal, Mexico, Brazil, and China.[3] An office was opened in London for the closer integration of the continental agencies and as a source of British and European market news. This network of domestic and overseas agencies constituted a vital feature of the pool mechanism for the disposal of Canadian wheat. The disbanding of these agencies and the discontinuance of direct selling were among the first steps taken by J. I. McFarland when he assumed the management of the Central Selling Agency in November, 1930, on appointment by the federal government at the request of the lending banks. McFarland took this action, as he said, "to demonstrate beyond possibility of doubt the truth or otherwise of the statement frequently made that the maintenance of direct representation overseas has militated against the sale of Canadian wheat."[4]

The concept of orderly marketing is one of the most elusive and most difficult of definition to be encountered in a study of co-operative marketing. Nevertheless the idea of orderly marketing had become an integral part of the rationale of the co-operative movement in the United States by the early 1920's, it was accepted as epitomizing the main marketing

[3]*Ibid.*, p. 273.
[4]H. S. Patton, "The Canadian Wheat Pools in Prosperity and Depression," from Norman E. Himes, ed., *Economics, Sociology and the Modern World: Essays in Honor of T. N. Carver* (Cambridge, Mass., 1935), p. 36.

objectives of the large-scale, centralized, pooling type of association which served as a model for the Canadian pools, and, as such, it found ready expression and acceptance among the grain growers of western Canada in the early inter-war years. Paradoxically, the elusiveness of the concept is rather to be welcomed than deplored, for while it renders definition difficult it imposes the requirement of a searching analysis which cannot fail to reveal certain at least of its inherent inconsistencies.

A student of American agricultural co-operation and co-operative development expressed himself on the subject in the late 1920's in part as follows:

Orderly marketing is often discussed but seldom defined. It is one of the popular catch phrases of the co-operative charlatan who holds forth the possibility of price control. It is also one of the ends sought by co-operative associations that base their programs upon the giving of efficient service. Orderly marketing is desired by primary producers, by transportation agencies, by wholesalers, retailers, and ultimate consumers. Not all who advocate orderly marketing have the same thing in mind, but the phrase is at least suggestive of an end that is greatly desired.

Orderly marketing has been defined as "the marketing of the commodity at the right time and place, in the right quantity and quality". Such a marketing program would eliminate gluts and famines. It would place every commodity on the market at the time and place that would give the greatest net return to the producer. . . .[5]

To define orderly marketing as the marketing of a commodity "in the right time and place, in the right quality and quantity" provides a concise but not very informative statement. To elaborate this definition by saying that such a marketing programme would remove gluts and famines and would yield the greatest net return to the producer begs many of the important questions at issue. In terms of price analysis it is obvious that such a principle of marketing would require quite different courses of action for competitive and monopolistic producers. For the competitive producer, market price is a given fact, quite independent of the marketing decision of any individual. For the monopolist that condition does not prevail, and market price is not independent of the producer's sales policy. The co-operative sales agency in the agricultural field ordinarily operates under market conditions which are neither purely competitive nor purely monopolistic. While such associations cannot direct their sales policies toward the maximization of net producer income according to the comparatively simple theoretical rules of

[5]H. C. Filley, *Cooperation in Agriculture* (New York, 1929), p. 411. Filley's quotation is from James E. Boyle, *Marketing of Agricultural Products* (New York, 1925).

monopoly pricing, neither can they ignore the fact that market price is no longer a given datum independent of their selling policy. The latter point is true, furthermore, whether the co-operative agency consciously pursues a sales policy in conformity with some orderly marketing formula or not.

With or without the possibility of precise definition, the concept of orderly marketing appeared to be made to order in terms of Canadian wheat marketing problems as interpreted by the producers in the early 1920's. Questions of quality of product and of choice among the various overseas markets were apparently of negligible importance. Questions of timing and quantity, on the other hand, appeared obviously to be at the root of the growers' marketing difficulties. Stewart and Riddell did not use the term, "orderly marketing," in the report of pooling possibilities which they presented to the Saskatchewan government in 1921. Nevertheless, their entire analysis of the relative merit of various kinds of pooling organizations was based upon a conviction of the glut-and-famine shortcomings of the open market system. Under the open market system, they said, ". . . wheat is daily offered for sale in quantities which reveal a lack of correspondence between actual supply and demand. During the early months of the season, when a large volume of wheat is offered for sale, there is a natural decline in prices. This system of competitive selling permits of no intelligent regulation of supplies of either the farmers' wheat or of that owned by the elevator companies. . . . The object of the Pool . . . would be the stabilization of prices by a more even distribution of supplies. . . ."[6] Later in their report they referred to the fact that under the open market system, upwards of three-quarters of the Canadian crop was "thrown on the market during a period of three months." Under a pool, however, "the movement of the crop would be more evenly spread over the whole crop year, thereby undoubtedly avoiding gluts of grain, and consequent depression of prices, which usually occur during the first three months of the crop movement."[7]

The extreme seasonal variation in Canadian wheat marketings and the apparent effects of this variation on the prices secured by producers for their wheat were matters of constant concern to Canadian growers in the early inter-war years. Producers who dumped their annual crop on the market within a fraction of the crop year—so it was argued— were obviously creating a glut of their product and could therefore expect nothing but unnecessarily low returns for their efforts. The specu-

[6]James Stewart and F. W. Riddell, *Report to the Government of Saskatchewan on Wheat Marketing* (Regina: King's Printer, 1921), p. 10.
[7]*Ibid.*, p. 15.

lator bought up the wheat at the depressed autumn price, the argument continued, held it till later in the crop year and extracted an indeterminate but substantial speculative profit for his shrewdness and perspicacity. The Wheat Board of 1919–20 had clearly marketed on a basis which assured the return of these speculative margins to the producer. A wheat pool of sufficient magnitude and with adequate resources should be able to achieve at least a significant measure of the success which had come so easily to the Board.

This may be taken as the gist of the argument propounded by Sapiro in the inspirational addresses which he gave in Alberta and Saskatchewan in August, 1923, on the eve of the inauguration of the provincial pools. By that time inspiration was of immeasurably greater importance than argument, for the line of reasoning was familiar to all and was accepted as axiomatic by the great majority of western growers. As Mr. Yates says, "It was not that Sapiro had anything new to propose, but rather that he gave the farmers enthusiasm and confidence. . . . Sapiro's great merit was that he refused to see the difficulties. . . ."[8] In his address at the meeting in Third Avenue Church in Saskatoon, a meeting which McPhail, first president of the Saskatchewan pool and of the Central Selling Agency, described as "wonderful," Sapiro said in part:

The central problem of co-operative marketing, the central problem of the farm is to try to stop dumping by the farmers. Every farmer in the world who sells as an individual is dumping his product, and breaking his own price by the dumping process. The fundamental thing is to stop the dumping of farm products, stop individual selling, stop local selling, and organize the commodity on such a plan that you can sell a great portion of that commodity from one office on a straight merchandising plan. By merchandising we mean control of the flow of any given commodity, so that it goes to the markets of the world, wherever they are, in such times, and in such quantities that they will be absorbed at a price that is fair under current conditions. Stop dumping and substitute merchandising. . . .[9]

By the time the pools were well embarked in the grain marketing business the pool antidote for dumping came increasingly to be identified as orderly marketing. A pamphlet published by the Saskatchewan pool in 1928 contained the following statement:

The Wheat Pools have introduced orderly marketing. This is no longer a myth, or a dream of things hoped for, but an accomplished fact. Farmers of Western Canada still deliver their wheat in the proportion of approxi-

[8]S. W. Yates, *The Saskatchewan Wheat Pool*, edited by Arthur S. Morton (Saskatoon: United Farmers of Canada, Saskatchewan Section, Limited, 1947), p. 46.

[9]*Ibid.*, pp. 78–9.

mately 75 per cent. in the fall months of the year, and 25 per cent. in the other eight or nine months; but no longer is it sold in that proportion. . . . Instead of 75 per cent. of Pool wheat being sold in the fall months of the year, approximately 40 per cent. is now sold during that period, which is in some ways the natural marketing period for the Canadian wheat crop.[10]

If the sale of an undue proportion of Canadian wheat in the months immediately following harvest constituted dumping, the corollary as it appeared to some people was that orderly marketing implied the sale of the Canadian crop in a uniform flow throughout the year, for example, in twelve equal monthly instalments. Pool officials found it necessary repeatedly to correct this impression. As suggested by the above quotation, officials of the pools felt that they had achieved a substantial measure of orderly marketing when they sold 40 rather than 75 per cent of the Canadian crop in the fall of the year which, they added, "is in some ways the natural marketing period for the Canadian wheat crop." At the International Wheat Pool conference in Kansas City in May, 1927, H. W. Wood, president of the Alberta Pool, stated:

"Orderly marketing" is a dangerous phrase. Too frequently it is taken to mean the systematic placing of wheat on the market in equal portions, day by day or month by month. This may be "orderly dumping", but it is not marketing in any sense of the word. The selling of wheat is a purely business proposition, and the Pool must decide when it is best to sell or hold, just as the buyers have to decide when to buy or not to buy.[11]

Pool officials recognized from the very early stages of the establishment of the pooling organization that the adoption of an orderly marketing programme provided no clear-cut formula for the guidance of their day-to-day sales policy. In November, 1926, when the Central Selling Agency had but recently completed its first year of operation, McPhail spoke publicly in support of heavy autumn selling by the pooling organization. "There is a bigger demand for Canadian wheat in the fall months than at any other season, and a much larger quantity can be disposed of now than in other months. You have to sell when the demand exists or you will find yourself holding the bag in your hands. We have already had that experience to some extent."[12]

Since it is quite clear that the adoption of an orderly marketing programme provides no specific guidance for the day-to-day sales policy of a pooling organization it becomes a matter of importance to observe

[10]*The Saskatchewan Co-operative Wheat Producers Limited: Its Aims, Origin, Operations and Progress, June 1924–January 1928*, Pamphlet no. 1 (Regina, 1928), p. 8.
[11]As cited in Patton, *Grain Growers' Coöperation in Western Canada*, p. 270.
[12]*Ibid.*

how the pools arrived at decisions governing actual sales. The wheat from the three provincial pools was, of course, all disposed of by the Central Selling Agency and its sales organization was under the direction of a general sales manager. This is the basis for the common claim that the pools substituted a single sales agency and selling authority for the 140,000 selling authorities which their membership constituted in pre-pool days.[13] When this single agency is pictured as acting without division of authority, in constant and immediate possession of every scrap of market information, and with consummate skill, it is easy to argue that pool selling must have constituted a tremendous advance over the disorganized and haphazard selling efforts which characterized the marketing activities of scores of thousands of individual farmers.

These ideal conditions were not, however, all fully in evidence. With their network of domestic and overseas agencies the pool obviously had unsurpassed access to market information. The ability of the sales staff is not in question. But to picture pool selling policy from day to day and from week to week as the product of the decisions of a single man or of a unified decision-making body is to misrepresent the situation as it existed. The general sales manager was responsible to, and in consultation at all times with, the board of directors of the Central Selling Agency. This board comprised three representatives from the directorates of each of the provincial pools who were responsible to their respective directorates and thus to their respective provincial membership groups. The actual decision regarding selling policy at any one time was therefore likely to represent at best a more or less satisfactory compromise among the diverse views of the sales manager and various members or sections of the central board.

Evidence on this point is for the most part scattered and elusive but a great deal is available in the published diary of A. J. McPhail.[14] Reference to that source is essential in order to illustrate, not the rightness or wrongness of any of the persons or groups involved but only the extreme difficulties encountered by the Central Selling Agency in formulating what may so readily be portrayed in the abstract as a "unified" selling policy.

[13]"The 'orderliness' of pool marketing, in short," said Dr. Patton, "lies not in any more measured movement of wheat from farm to consuming market, but in unified selling on the basis of available knowledge of world supply and demand conditions" (*ibid.*, p. 271).

[14]H. A. Innis, ed., *The Diary of Alexander James McPhail* (Toronto, 1940), hereafter cited as *McPhail Diary*. The interchange that follows between officers of the Central Selling Agency will be cited in the text as page references to this diary.

Sharp differences of opinion on selling policy were evident in the deliberations of the directorate of the Central Selling Agency from the very start. Organized in August, 1924, this agency was headed by A. J. McPhail as president, H. W. Wood as vice-president, and C. H. Burnell as secretary, these three being chairmen of the Saskatchewan, Alberta, and Manitoba provincial pools respectively. D. L. Smith and C. N. Elliott, former officials of the United Grain Growers, were appointed eastern and western sales managers respectively. On November 3, 1924, Smith wrote McPhail a letter, a brief excerpt from which clearly indicates the divergence of views between the president and vice-president in regard to sales policy:

I am to a certain extent in sympathy with your views in regard to the method we should adopt in the marketing of our wheat, but it must also be remembered that it is our duty to see that the farmer secures as much as possible for his wheat. Therefore as Mr. Wood stated the other day it is not sufficient to only get a good average price if we feel in our own minds that we can do better. I have always been a little backward about following the policy as outlined by Mr. Wood but I think if our position was sufficiently secure that it would be good trading on our part to occasionally buy on sharp breaks to the market. (p. 114)

The divergence of views which was clearly evident in the first weeks of the Central Selling Agency's operations developed into a major crisis and brought it to the verge of collapse before the end of its first crop year. In December, 1924, Smith was sent to England, and his earliest reports predicted a poor demand for Canadian wheat. Toward the end of January, 1925, however, he cabled: "Tell directors I believe see much higher prices. Europe requirements large, hardly be enough wheat to go around. Advise selling on moderate scale." (p. 115) This information failed to impress McPhail, who continued to urge the steady disposal of existing stocks of wheat. A day or so after receiving Smith's cable he wrote to Burnell urging continuous selling, and saying that, otherwise, "if . . . there was to be a slump we would be justly accused of, to say the least, mighty poor judgment" (p. 115). At the next meeting of the central board, however, on January 30 and 31, 1925, the view was strongly pressed that the pool should not only withhold the wheat it had but should also go into the futures market and purchase May wheat. McPhail's diary summarizes the dispute:

Mr. Wood is insistent on getting long May and I am as insistent we shall not. *January 31.* We had a very tense discussion on selling policy during the afternoon. Wood tried to force his theory of getting long May. I fought it and would have resigned if they had insisted on putting it into practice. I had to agree for the sake of harmony to sell no more May options, which I feel is bad policy in view of the present high speculation market. Wood withdrew the resolution. (p. 115)

On February 23 McPhail entered in his diary a general statement on marketing policy:

I have always taken the stand that a pooling organization should pursue a policy of trying to get an average price for the season. To the extent that it departs from a policy of that kind by trying to get more than an average price to that extent does it place itself in a position of getting something less. We must above everything else pursue a steady selling policy to eliminate the chances of speculation as much as possible in the marketing of our wheat. (pp. 115–16)

Selling policy was a bone of contention at every meeting of the Central Selling Agency's board of directors throughout the winter months of 1924–5. The clash over May futures was settled by a compromise which was distasteful to both contending groups. On February 16 the board had a "long discussion on selling policy. . . . Wood and the Alberta men reversed their stand on selling policy to that of the last meeting." (pp. 115–16) On March 6 they "discussed selling policy at length." On the seventh they had "another discussion on selling policy in the afternoon. Wood would not sell a bu. of wheat under $2.00. It is $1.88 today. Absolute nonsense and exceedingly dangerous. . . . Wood would like to buy options at these prices and make a profit when wheat goes up, as he is sure it will." On March 9 McPhail wrote to Burnell:

I feel that we are in a very dangerous position having a section of the Board determined to pursue a selling policy with the definite opinion that prices are going to much higher levels. As I have said on many occasions if we do not consider present prices very attractive from the standpoint of the farmer and if we do not show that we consider them so by reasonably liberal selling if there should by any chance be a permanent drop in the market it would seriously affect the pool for many years to come. (p. 116)

During March the pool declared an interim payment of 35 cents per bushel which, together with the initial payment of $1.00, committed them to a net return of $1.35, basis No. 1 Northern, Fort William. On March 31 a slump occurred, with the May future falling to $1.41¾ and closing at $1.48½. On April 3 May wheat dropped to $1.36 and closed at $1.38⅞. It was current rumour that the slump was caused by a deliberate bear raid, participated in by the Saskatchewan Co-operative Elevator Company, with the intention of smashing the pool. On March 31 McPhail had written: "Smith and McIntyre are having a very anxious time. . . . If the pool is smashed there is only one man to blame. . . . Even now I cannot see with the knowledge I had, how I could have done more to avert this near calamity." (p. 117) And on April 2: "We had a discussion on selling policy. There is now no difference of opinion on the Board regarding the necessity of making ourselves safe by selling as quickly as we can without further demoralizing the market. We are still in a strong position and the banks have not bothered us. . . ."

There was increasingly strong sentiment in favour of counter action by the pool to give an upward lead to the market. This was finally agreed upon. McPhail wrote on April 4:

We decided we would have to take drastic measures if wheat was to go lower. According to reports the market was likely to go down today. We simply cannot stand by and let the present situation continue or grow worse without putting up some kind of a fight. . . . We must keep a 10 cent margin above the initial [plus interim] payment. This morning we had only a 3 cent margin on 1 Northern on a basis of $1.35 and our margin on lower grades has been wiped out several days ago. . . . Market closed at 1.44½. We took steps to help it today in the hopes it would strengthen and enable us to sell wheat for export over the week-end. . . . (p. 118)

On May 6 he wrote:

May closed 1.46¾ or 2½ higher than Saturday's close. We bought 500,000 on the Winnipeg market and sold it on Chicago. . . . We are simply forced to take these measures to fight the bears on the market. Apparently a few strong grain interests can bear down the market if there is no bull resistance. All the bullish interests are afraid to buy for they do not know when the pool may be forced to unload. The pool appears to be the only organization that can go in and change the trend of the market and to do it we must take steps which we would not under ordinary circumstances take. But we must fight the devil with his own weapons. (p. 118)

Whatever the forces at work and whatever the degree of efficacy of the pool's corrective action, the crisis was quickly ended. Prices rose and the pool's margin with the banks was restored. Between April 20 and 22 McPhail wrote in Winnipeg:

We had an executive meeting all day. Discussed many matters but mostly selling policy. We don't see for the moment how the different views can be reconciled. I think a good price is good selling. Mr. Wood thinks we must use our judgment and do good selling whatever that is. I hope Mr. Wood can see his way clear to agree to a practical and common-sense selling policy. *April 21.* Wood, Burnell, and Brownlee and I spent all day again in meeting. It was very difficult sometimes to keep from saying some things which I suppose would be better left unsaid. Specially in view of the terrible position the pool was very nearly placed in recently by ———'s theoretical nonsense. Wood is a very likeable man apart from his philosophizing about practical business matters. *April 22.* Talked with Brownlee between 6 and 7 p.m. We agreed everything possible must be done to keep the three pools together. (p. 119)

The immediate crisis over, McPhail nevertheless continued to express concern over the quantities of pool wheat which remained unsold. On June 22, 1925, he wrote to Smith, "I do not feel very easy about the large quantity of wheat which we still have on hand. I hope the prices will keep steady." (p. 120) And on June 30 he wrote, "We are in a deuce of a fix having a lot of wheat and no demand." Information on the size of the wheat pool's carryover at the end of the 1924–5 crop year is not at hand. The pools received 81.7 million bushels of wheat during the year and on August 13, 1925, decided upon a final payment which brought the total net return for the year to $1.66 per bushel, basis No. 1 Northern, Fort William (p. 120). This was a considerable advance over the corresponding price of $1.00 per bushel for the preceding crop year and critics were therefore not likely to be either numerous or active. Nevertheless, Patton records from "personal information" the admission of pool officials that too large a proportion of the 1924 crop had been held back, if not beyond the end of the crop year, at least until unduly late in the season when a variety of influences were working in the direction of lower prices.[15]

Differences of opinion among members of the central board concerning selling policy were a basic and persistent impediment to the formulation and pursuit of unified objectives. Of lesser but by no means negligible importance were managerial difficulties involving the relationship of the directorate to salaried officials and the co-ordination of the latter among themselves. Almost immediately after his appointment as general sales agent, Smith wrote McPhail a letter, cited above,[16] in which he clearly displayed vacillation between acceptance of the general sales policy of McPhail and sympathy with the divergent views of Wood. The establishment of two sales agencies in the prairie provinces, accepted as necessary because of provincial claims and the opening of the western grain trade route in the inter-war years, intensified the difficulties of formulating a unified selling policy. On January 7, 1925, Smith wrote to McPhail:

I think the time has come when it should be made absolutely clear to McIvor [the sales agent in Calgary] that there must be someone to act as sales manager; it is a foregone conclusion that we are going to run into a lot of trouble if we attempt to handle our selling through two distinctly separate offices. After all, the Winnipeg office sells about 95 per cent of the pool grain as Calgary makes most of their sales basis, the option, so that it is up to Winnipeg to decide when to sell out the options. P.S. It seems the Calgary office are under the impression they are working for the Alberta pool instead of the Can. Co-op. (p. 113)

[15]*Grain Growers' Coöperation in Western Canada*, p. 270.
[16]P. 227.

Months later, in August, 1925, McPhail was in Calgary inquiring about the opening of a pool office in Vancouver, which had taken place without the knowledge of the board. Wood "said he didn't know anything other than that McIvor had spoken to him about it and he thought it was all right. McIvor said he talked it over with Wood and he thought the Board knew about it." McPhail asked McIvor to come to Winnipeg "in order to have a full discussion re relationship of Calgary and Winnipeg office." A few days later McPhail "had a long talk with McIvor on co-operation."

In the succeeding year or so McPhail expressed dissatisfaction repeatedly with the progress made by the sales staff in disposing of wheat. On November 12, 1926, McPhail and the board "discussed selling policy with Smith in the afternoon. I am not satisfied that he has sold as much as he should have this last month."[17] A month later the Board "discussed the question of getting a new chief salesman." In February, 1927, McPhail was "not at all satisfied with our market position. We have not enough wheat sold." On March 8 he was "better satisfied with our selling position than [he had] been this year so far." On March 10 the Board "spent all afternoon with Smith and Folliott. ———— as usual wasted a great deal of the time trying to convince the Board it should not sell wheat or at least sell much more slowly." Smith resigned in the autumn of 1927 and was replaced as general sales agent by George McIvor from the Calgary office. Smith was appointed to take charge of the London office. On September 19, 1927, McPhail wrote concerning the change: "Mr. McIvor impressed the Board as the best real co-ordinator and co-operator on our selling staff. Mr. Smith is a man of real ability but he has not shown the disposition to co-operate that the Board feels he should have shown in the past. . . . It is essential that the Board retain a firm hold over the whole organization and everyone in its employ from the Manager down." (pp. 171–2)

Divergence of views on how an orderly marketing programme should be made effective in day-to-day selling policy stemmed from a fundamental difference of opinion about the possibilities of agricultural price control. The opinions expressed varied widely in meaning and shades of meaning, and classification may contain an element of unfairness because of over-simplification; nevertheless, something is to be gained by an effort to outline the leading views on agricultural price control which prevailed in the wheat growing regions in the 1920's. The more moderate view was the one adhered to with substantial consistency by A. J. McPhail and his considerable group of followers. This view was

17*McPhail Diary*, p. 155.

that a powerful, centralized, co-operative marketing agency, such as the wheat pools, might effect a measure of inter-seasonal price stabilization within any one crop year and might secure for its members the full average world price less minimum marketing costs for the year in question, but the world price itself would inevitably remain beyond the influence of any single co-operative group.[18] Orderly marketing could be made effective only within each crop year and only in conformity with world supply and demand as they currently existed. The average price as interpreted in this context would, in theoretical terms, be an equilibrium price, the highest price which would clear the market of one year's production before the next year's crop came forward.

The other view was considerably more ambitious. It started from the assumption that the agricultural producer was entitled to prices which, although variable, could nevertheless be defined at any point of time as fair in relation to the prices obtaining in other sectors of the economy.[19] Its further assumption was that such prices could be assured by organization and marketing control and that if existing instruments of control were inadequate, it was the opportunity and responsibility of agricultural producers to fashion instruments which would *not* prove inadequate. A branch of American agricultural thought was influential in developing this view in the Canadian West. The possibilities of success

[18]A previous quotation from McPhail (*Diary*, pp. 115–16) included a statement which he made in 1925: "I have always taken the stand that a pooling organization should pursue a policy of trying to get an average price for the season." In his evidence before the Committee on Agriculture and Forestry of the United States Senate in 1929, McPhail stated in regard to pool selling policy, "we try to stay off the market when it is in a weak condition and try to take advantage of the market when it is strong due to a great many factors. . . . I am coming more and more to the belief that in the final analysis it is the actual amount of wheat that is available in the world from time to time that has more to do with determining the average price level of wheat the same as any other commodity." (p. 196 n.) And before the Empire Parliamentary Association in London in 1930, "We have no illusions regarding the power of an organization such as ours controlling as it does 50 percent of Canadian wheat to influence prices unduly. I think that quite impossible because, after all, whether you have a high level of prices this year or a low level of prices next year depends more than anything else on the abundance or shortage of supplies. . . . I believe in some years the pool's operations have resulted in a stabilized upward level of prices. . . . at certain times an organization such as the pool can have a steadying and stabilizing influence on the market by not pressing when there is no demand." (p. 205 n.)

[19]This view can readily be recognized as the one which, by the early 1920's, emerged in the United States as the concept of "fair exchange value" and, a few years later, appeared as the much more pervasive and persistent philosophy of "parity price." For an outline of the development of these ideas in the United States see J. D. Black, *Agricultural Reform in the United States* (New York, 1929), particularly chaps. viii–x.

in achieving its objectives were repeatedly supported by appealing to the analogy of price control in the industrial and to a lesser extent in the labour sectors of the modern economy.[20]

The conflict between these views on agricultural price policy permeated the agrarian movement in western Canada throughout the inter-war years. It occupied a central place in the deliberations of the farmers' educational organizations and, as analysed briefly above, created endless friction among those immediately responsible for the formulation of policy in the leading commercial organizations, the pools. At times it flared into open dispute, at other times it smouldered menacingly beneath the surface of nominal agrarian unity. The decision to organize the pools on a voluntary contract basis represented a temporary fusion of divergent viewpoints but pool selling policy over the succeeding years was formulated only at the cost of endless painful compromise.

The more radical view of agricultural price policy first found formal expression in the West in the Farmers' Union of Canada which was organized at Kelvington, Saskatchewan, in 1921. Earlier in the year a group of farmers at Ituna, Saskatchewan, had formed an organization "with the object in view of supporting and affiliating with farmers' organizations in all the large producing countries to obtain control of all main farm produce, *to regulate and obtain reasonable prices above cost of production,* and also to protect the farmers' interests by the support and strength of their own organization."[21] On July 1 the secretary and another member met with representatives of another new organization formed at Kelvington, the Industrial Farmers' Union of Canada. The two groups merged to form the Farmers' Union of Canada, an organization based essentially on a distrust of the conservatism of the leadership of the Saskatchewan Grain Growers' Association and on zeal in the cause of agricultural price control. Mr. L. C. Brouillette was

[20]H. W. Wood, speaking at the International Wheat Pool Conference held at St. Paul in February, 1926, deplored the dumping of wheat on world markets and said, "If this wheat was sold intelligently, systematically and fed to the consumptive demand just as that demand developed, *we could maintain the price of our wheat on a level with the prices we have to pay* and we would not need any legislation to assist us in doing that either. . . ." (as cited in *McPhail Diary,* p. 129 n., italics added). For a contemporary rationalization of wheat pool selling policy with constant reliance on the analogy of the business world and with the use of such terms as "modern business methods," "tested principles of modern salesmanship," "scientific selling," "merchandising," and much of the rest of the jargon of the success-and-progress-inspired urban business community see Walter P. Davisson, *Pooling Wheat in Canada* (Ottawa, 1927), *passim.*

[21]Quoted in Yates, *Saskatchewan Wheat Pool,* p. 57 (italics added).

one of the first members of the Farmers' Union; he was elected provincial organizer at the 1921 convention and vice-president a year later.[22] As vice-president of the Saskatchewan wheat pool throughout the years of contract pool activity (he became president in 1931 on the death of McPhail) and as a director of the Central Selling Agency, he was a persistent advocate of price control, an objective which was regarded as wholly unrealistic by some at least of his colleagues on these boards. As vice-president of the Farmers' Union he was among the first to urge that Sapiro be invited to Saskatchewan to outline the co-operative pool marketing system so much in vogue in California and other parts of the United States. He maintained contact with Sapiro throughout the years of pooling activity and enlisted his support in matters which came to serious issue in the directorates and membership of the pools. Chief among these particular issues were the elevator dispute and the conflict over compulsory pooling.[23]

The decision to organize and conduct the marketing of wheat by means of a voluntary contract pool did not remain unchallenged in Saskatchewan throughout even the first five-year contract period. It was a question of price policy and of price control possibilities. We have already noted the price control objectives enunciated by the organizers of the Farmers' Union in 1921. The dominant group in that organization regarded and continued to regard their statement of objectives as far removed from empty verbiage. The institution of the pools, which the group effected to a considerable extent through their own initiative and drive, satisfied their anticipations for a year or so as being at least a step in the right direction. A membership drive aimed at securing growers' signatures to patronage contracts on a voluntary basis had a potential upper limit of 100 per cent and the maximum possibilities of price control on a national basis could require no less: anything short of complete participation suggested obvious weakness. Moreover, with voluntary contractual participation considerably short of completeness, an inability to control prices, even though it might be due to the most fundamental of economic considerations, could nevertheless be attributed to the gap between actual and complete participation on the part of the producers.

The achievements of the pools during the first two or three years of operation were in many ways spectacular but in certain important aspects

[22]*McPhail Diary*, p. 46.

[23]For indications of the significance of the interest of Sapiro in the pools and of the part he played in relation to the internal controversies over policy see *ibid.*, pp. 50 ff., 88–90, 158–9 and *passim*; also Yates, *Saskatchewan Wheat Pool*, chap. VII and ff.

their accomplishments could be regarded as only reasonably satisfactory. The Central Selling Agency had skirted the margins of disaster in the spring of 1925 and its best efforts had not served to avert a decline in the price of wheat from approximately $2.00 per bushel, basis No. 1 Northern, Fort William, in the early months of 1925, to a low of $1.38⅜ in April (with an average of $1.56 for the month of April). The net yield of pooled wheat fell from $1.66 per bushel in 1924–5, basis No. 1 Northern, Fort William, to $1.45 and $1.42 for the crop years 1925–6 and 1926–7 respectively.[24] These prices were easily justifiable on the ground that they represented good average prices for each of the respective crop years in question. Significantly, however, pool officials found it necessary to offer justification in answer to considerable vocal criticism. In retrospect it is fair to say that the achievements of the pools in the early years were such as should have satisfied all reasonable expectations. At the time, however, the expectations of pool members were by no means entirely reasonable. In any event, the expectation assiduously fostered by an influential element of agrarian leadership throughout the 1920's, that a voluntary contract pool might exercise a significant degree of control over the price of wheat, had, by the observable experience of the Canadian wheat pools in their early years, proved to be chimerical.

Whatever might be the basic inadequacies of the pooling method to achieve price control, the most obvious weakness of the Canadian wheat pools in their early years appeared to lie much nearer to the surface and to be completely and readily removable. This apparent weakness was that voluntary contractual patronage fell far short of total Canadian wheat marketings. Starting at 37 per cent for the 1924–5 crop year, deliveries to the Central Selling Agency ranged from 51 to 53 per cent of total western deliveries over the next five years.[25] The provincial pools by 1928–9 had approximately 140,000 members under contract. These represented approximately one-half of the total number of farmers in the prairie provinces.[26] The pools constituted, in effect, an important alternative to, rather than a replacement for, the open market system. The possibilities of price control might well be negligible for an organization in such a position. If pooling were to be given a serious trial in the regulation of wheat prices, it would be necessary, so it was argued with apparent cogency, that the other half of the wheat crop should be

[24]For a summary of pool price experience see Table IX in chap. XIII below.
[25]See *ibid.*
[26]Census returns indicated 248.2 thousand farms in the prairie provinces in 1926 and 288.1 thousand in 1931.

brought within the control of the Central Selling Agency so that it, like
the Canadian Wheat Board of 1919–20, should have exclusive disposal
of the output of western Canada.

The movement for a compulsory 100 per cent pool did not achieve
the same degree of momentum in all the prairie provinces. It was particularly strong in Saskatchewan. By the end of 1926, pool officials
estimated that 80 per cent of the acreage ordinarily seeded to wheat
in the province was under contract to the pool.[27] An anomoly apparently
without completely satisfactory explanation in view of the estimated
contractual coverage was that the Saskatchewan pool secured only
approximately 56 per cent of the annual wheat deliveries in the province.
On either basis there was ample room for extending the pool's control
beyond that which had been achieved on a voluntary basis. By mid-
summer of 1926 the Farmers' Union of Canada and the Saskatchewan
Grain Growers' Association had amalgamated under the name of the
United Farmers of Canada (Saskatchewan Section) Limited. The advocates of agricultural price control who had comprised the active leadership of the Farmers' Union remained prominent in the new association
and through it they exercised influence and pressure toward the establishment of a compulsory pool. The campaign for the compulsory
pooling of wheat—of all grain, in fact—began in Saskatchewan in 1927
and continued with increasing power over a four-year period until the
denouement of judicial defeat in 1931. It constituted an intensively
divisive force throughout both the United Farmers of Canada and the
Saskatchewan wheat pool.

Speaking at a number of meetings in Saskatchewan in the summer of
1927, Sapiro advocated a compulsory pool in terms which were known
thereafter as the "Sapiro plan." A quotation from one of his addresses
will indicate the position he took:

I believe that when two-thirds of the acreage, or two-thirds of the growers
in this province do what they have done it certainly means that the great
body of agricultural workers in the province want co-operative marketing.
When that has been demonstrated, I believe the Saskatchewan Legislature
should pass a law providing that such membership and such contracts
represent the overwhelming desire of the growers of this province and that,
therefore, all growers alike should be compelled to deliver all grains to the
Pool only, and that the Pool shall sell all the grains produced in this
province. . . . [This is] the practice of majority rule, which is carried out
in all democratic countries.[28]

[27]Saskatchewan Co-operative Wheat Producers Limited, *Second Annual Report*
(Regina, 1926), p. 3. The figure given was 79 per cent as of the date of the
Report, Oct. 12, 1926.
[28]Yates, *Saskatchewan Wheat Pool*, pp. 137–8

After Sapiro's first public advocacy of compulsory pooling the issue agitated the country-side. It was raised and fought over in the meetings of the United Farmers of Canada and of the Saskatchewan wheat pool delegates. It was the subject of special consideration at meetings between executive officers of the U.F.C. and of the wheat pool directorate. On the eve of the annual meeting of the Saskatchewan wheat pool delegates in November, 1927, McPhail wrote: "So far as the Sapiro plan of making the pool compulsory after we secure 75 or 80 per cent of the crop in this province is concerned I think the less said about the scheme the better."[29] The delegates listened to a debate by four of their number on the topic, "Resolved that we are in favour of a 100 per cent pool on the Sapiro plan."[30] The group was far from ready to endorse the plan and, after discussion, referred the matter back to the educational organization, the U.F.C., with the resolution "that the United Farmers of Canada, Saskatchewan Section, as an educational organization representing Saskatchewan agriculturists, use their educational facilities to obtain a thorough discussion of compulsory pooling throughout the province, so that such knowledge and conclusion obtained may be used as a guide to the desirability of the compulsory pooling system."[31] In December the executive of the U.F.C. met with the directorate of the Saskatchewan pool. McPhail wrote: "*December 13, 1927.* Our Board met the executive of the United Farmers after lunch. Stoneman, Edwards, Mrs. Hollis, Bickerton, Williams, Murray, and Thrasher were present. Stoneman was quite anxious to have the Wheat Pool Board committed to a compulsory 100 per cent pool but I made it quite clear that . . . I think the idea is out of the question when mentioned in the same day [*sic*] as co-operation."[32]

In February, 1928, the question was raised for consideration at the annual convention of the U.F.C. By a narrow margin the delegates adopted the resolution: "That we go on record as being in favour of having the provincial legislature enact a law which will make it compulsory for every wheat grower and farmer to market his or her wheat through the Wheat Pool when 75 per cent of the farmers in Saskatchewan sign the Wheat Pool contract."[33] The delegates to the pool meeting in November agreed that they, in co-operation with the United Farmers of Canada, should attempt to determine the general opinion of growers on the question of compulsory pooling and report back to

[29]*McPhail Diary*, p. 146.
[30]*Ibid.*, p. 147.
[31]Yates, *Saskatchewan Wheat Pool*, p. 139.
[32]*McPhail Diary*, p. 185.
[33]Yates, *Saskatchewan Wheat Pool*, p. 139.

the June meeting of the pool delegates. McPhail wrote in his diary, November 28: "Meeting [of delegates] ended a little after 6 P.M. I was greatly disappointed wtih the apparent support in the meeting for the compulsory pool. It will kill the organization if it becomes the policy of the pool to seek compulsory legislation."[34]

By 1929 the campaign for compulsory pooling in Saskatchewan had become intense. In February the U.F.C. fought over the matter in convention and adopted a resolution instructing the executive to inaugurate an educational programme to influence public opinion in the province in favour of compulsion. The reasoning of the resolution was that the partial control exercised by the pool during the first contract period had permitted a certain increase in prices but that only complete control of grain marketings could secure for the producer the full value of his product. The approval of this plan of action, along with other elements of incompatibility, led shortly after to the resignation of five of the officials of the U.F.C.[35]

In the same month, McPhail, in Chicago, met Burnell returning from New York, and the latter reported that Sapiro was coming to Saskatchewan to "put over" the compulsory pool.[36] At the June meeting of the Saskatchewan wheat pool delegates, the committee on resolutions dealt with sixty-six resolutions relating to compulsory pooling, the resolutions, according to the chairman of the committee, being almost equally divided for and against the proposal.[37] The divisive nature of the compulsory pool question at this time is suggested by Yates, an uncritical

[34]*McPhail Diary*, p. 187.
[35]Yates, *Saskatchewan Wheat Pool*, pp. 139–40.
[36]*McPhail Diary*, p. 187.
[37]Yates, *Saskatchewan Wheat Pool*, p. 141. Describing the discussions of the compulsory pool question, McPhail wrote in 1929: *June 20.* "We [delegates] took up all the afternoon in a discussion of compulsory pool. ——— spoke for an hour and ten minutes and made quite an appeal of a kind. Very little argument.... *June 21.* Spent all day in discussing compulsion. Some good speeches. The vote on the amendment to approve of the plan and go out and educate for support lost by 94 to 56 counting my vote against. We did not count the vote on the main resolution which was overwhelmingly defeated. . . . It was a very satisfactory result and decides for a time this most dangerous question. *June 22.* It will be interesting to see how Sapiro will be received next week when he comes to advocate what the delegates have turned down. I hope most people will awaken to just what he is. *June 26.* Sapiro raked the pool Board fore and aft at the banquet and Arthur Meighen spanked Sapiro very courteously. *June 27.* They all seem to think that Sapiro hurt himself last night by his petty personal references." (*McPhail Diary*, p. 189) In December, 1928, McPhail wrote to Professor C. R. Fay, "In my opinion when you introduce compulsion you eliminate co-operation and I feel for many reasons that if this organization attempts to secure such legislation it will give the co-operative movement a setback for many years." (*Ibid.*, p. 185 n.) In December, 1929, McPhail wrote to the Hon. Mr. Doherty,

admirer of Sapiro. Speaking of Sapiro's visit to Saskatchewan to "put over" the compulsory pool, Yates sketches the situation in the following words:

In the month of June, 1929, for the fifth time, the prophet appeared, this time under the auspices of the United Farmers of Canada, to find the work of four years in the cause of unity in ruins at his feet. Indeed, it is not too much to say that the entire province was at this time a camp of warring factions. On one side the leaders of the U.F.C. and the more advanced of their members, assisted by Mr. Brouillette and one or two other members of the Pool Board, and on the other the President, Mr. McPhail, and a great majority of the members of the Pool Board; while the members of the Pool were apparently equally divided.[38]

Resolutions favouring legislation to establish a compulsory pool "so that speculation, manipulation and waste be eliminated in marketing all grain grown in Saskatchewan" were, however, defeated by a substantial majority of the delegates following a debate which lasted for a day and a half. Following this action the *Western Producer*, which heretofore had been used as an organ of publicity in favour of the campaign, withdrew its support. The rejection of compulsory pooling by the pool delegates did nothing to diminish the efforts of the advocates of the plan, whether in the province at large, in the directorate of the Saskatchewan wheat pool itself, or in the counsels of the U.F.C., where its chief advocates included the president, vice-president, secretary, and director of publicity (George H. Williams, A. J. Macauley, Frank Eliason, and George F. Stirling, respectively).[39]

On September 4 McPhail wrote in his diary that he had had "a talk with Brouillette about compulsion. He denied having in mind compulsory legislation that would not give the Government or the non-pooler representation on a selling board. Nevertheless he and his associates have been advocating just that. A most fantastic and silly idea." Later entries during the same month indicate the persistent friction within the Board.

September 7. Moffat, Fry and I had a long talk about the present serious situation in the province over compulsion. We agree that the time for aggressive action has come to combat the insidious campaign being carried on by the U.F.C. which is being directed by a Wheat Pool Board member against the policy of the Wheat Pool organization. A very peculiar and

---

"It is altogether too bad that all the energy that is being expended in connection with the advocacy of compulsion could not be used in furthering the interests of our co-operative organization along truly co-operative lines." (*Ibid.*) Also see *ibid.*, pp. 247–8 n.

[38]Yates, *Saskatchewan Wheat Pool*, p. 140.

[39]*Ibid.*, p. 141.

serious situation. *September 16.* ———— has evidently been attacking me personally in the country. I am surprised to hear it as I thought he was more loyal to associates. *September 20.* We had quite a long talk in the Board on compulsion. I expressed my views of the present situation as clearly as I could. It is very serious. *September 25.* Called at Hamilton's for lunch and stayed there until 5 P.M. talking mostly of compulsion and some way of avoiding a war. Will the U.F.C. take this disruptive question out of the pool and advocate if they want to, Government control of non-pool wheat through a Government agency? Let the pool develop as a co-op organization. (pp. 248–9)

McPhail campaigned throughout the province in the fall of 1929 against the compulsory pool plan. The pool delegates in November rejected resolutions calling for an educational campaign and a plebiscite. The executives of the pool and of the U.F.C. discussed the matter on different occasions. Feeling at the meetings ran high and recrimination was not lacking. By March, 1930, both groups were shifting ground and were less sharply opposed. On March 21 McPhail wrote in his diary: "Our Board was in session all day with the U.F.C. executive discussing the question of a compulsory pool. The U.F.C. officials have shifted their ground a great deal from the Sapiro proposals. Although I am opposed to any compulsion, their present plan is much more palatable and feasible than their first. It at least acknowledges the right of the individual to a voice in directing the affairs of the organization to which he would be compelled to deliver his wheat." (pp. 250–2) The meeting referred to in this citation was, however, one of the stormiest that took place between the groups at any time.

In June, 1930, the pool delegates defeated a motion to request the government for compulsory pooling legislation, but passed resolutions approving in principle 100 per cent pooling by legislation and requiring a referendum of the matter to all signers of pool contracts. McPhail wrote on June 3:

I think it is the proper thing now to take a vote of our members on the question and later a vote if necessary of all growers as to whether all growers will have to market their grain through a pool. I think probably it is necessary, not because I think it is right, but because of the kind of men who have been persistently advocating compulsion and the situation they have created through preaching that we will never get another sign-up and turning aside the minds of the people from co-operation. We may find it very difficult to get another sign-up but the advocates of compulsion are the real cause of that condition if it really exists. I really find it difficult to decide whether to remain with the organization or not. I think it quite possible we are done and yet it may not be so. I have no confidence whatever in some of the chief advocates of the scheme. . . . If I consulted my own desire only, I would resign tomorrow. (p. 253)

On July 2 he wrote to D. L. Smith:

I have not changed my mind on the question of the value of compulsory legislation. This proposal, however, is so very much different from the one that was being advocated by Sapiro and his admirers for over two years that the people generally are taking a different attitude towards it. It is at least democratic in that all growers, when compelled to deliver grain, will have a voice in directing the affairs of the organization. The really vicious principles in the former proposals are eliminated. (p. 253 n.)

The directorate proceeded with a referendum of pool members as called for by the delegates in June.[40] By mid-August the market situation had become desperate. The initial payment for the new crop had been reduced from 70 to 60 cents per bushel, basis No. 1 Northern, Fort William (later to be cut to 55 cents and finally to 50 cents per bushel), and contract signers were "bootlegging" substantial amounts of wheat outside the pool. The results of the referendum were not yet known when the executive of the pool went with the president of the U.F.C. to ask the Saskatchewan government for the compulsory pool for the crop then being marketed. The government gave no commitment. In September, after the results of the referendum of its members had been tabulated, the pool again requested legislation to provide for compulsory pooling under the exclusive control of the growers, and for a second referendum of *all* the Saskatchewan farmers to provide authority therefor. After consultation with its members the government declined to act on the request, but on December 30 Premier Anderson promised compulsory pool legislation at the ensuing session, the implementation of the plan to be contingent upon the outcome of a referendum of all the farmers within the province.

The assurance given by the government that it would make legislative provision for compulsory pooling brought to a sharp focus the opposition to the plan. By this time, however, the opposition was centred outside

[40]The question submitted to the contract signers was the following: "Are you in favour of your Directors asking the Government to pass a Grain Marketing Act to provide that all grain grown in Saskatchewan must be marketed through one Pool, provided, (a) That before the proposed Act should come into force it must receive a two-third majority vote in a special referendum of all grain growers in the Province, to be conducted by the Government; (b) That the Grain Pool to be provided for must be entirely under the control of the growers delivering grain, and further that all producers of grain (whether Pool or Non-Pool) must have an equal voice in the control of the organization?" See Saskatchewan Co-operative Wheat Producers Limited, *Sixth Annual Report* (Regina, 1930), p. 28. Fifty-nine per cent of the 82.9 thousand ballots sent out were returned. Of these, 34.6 thousand favoured the proposal for a compulsory pool, and 13.8 thousand voted against it. This was a favourable vote of 71.4 per cent of those who voted, or 41 per cent of the total membership. *Ibid.*

the wheat pool directorate and largely outside its membership. Late in January, 1931, a deputation of farmers presented a memorandum to the provincial Premier opposing compulsory pooling, and within a few days the same group formed the Association Opposing 100 Per Cent Compulsory Pool. The managing director of the Saskatchewan pool, R. J. Moffat, replied in a statement to the press and the statement was mimeographed for distribution under the pool's letterhead.[41] A few days later McPhail issued a statement in support of the legislation. "Personally," he said, "I am absolutely in favour of the proposal as contained in the Grain Marketing Act, as it will place Canadian wheat producers in a position to effectively co-operate with similar control bodies in other wheat exporting countries, in order to dispose of world wheat surpluses on a rational basis. . . . The whole world, in connection with all commodities, is undoubtedly moving steadily, and in some cases very rapidly, in the direction of national and international control."[42]

In February the Referendum and Grain Marketing bills were enacted and were assented to on March 11, 1931.[43] Within a month the Grain Marketing Act had been brought before the Saskatchewan Court of Appeal for a test of its constitutionality. On April 27 the four judges handed down separate written judgments which unanimously declared the legislation to be *ultra vires* the provincial legislature. The pool sought permission from the court to appeal the decision to the Privy Council. Meanwhile, W. A. Scott, a pool member against whom the pool had taken action for breach of contract, sought and secured a temporary injunction restraining the pool from using its funds for the purpose of appeal. The U.F.C. demanded that the provincial government either request the federal government to provide enabling legislation to make the Grain Marketing Act operative or, failing that, that the provincial government finance an appeal to the Privy Council. The government stated that it would do neither. The compulsory pool campaign was thus ended by July, 1931. By that time the pools no longer controlled the Central Selling Agency and wheat marketing policy was in the hands of the federal government.

41See "Pool Managing Director Replies to Anti Grain Marketing Act Memorandum" (Regina, January 29, 1931).

42Yates, *Saskatchewan Wheat Pool*, p. 150.

43*Statutes of Saskatchewan*, 21 Geo. V, c. 87, The Grain Marketing Act, 1931; and *ibid.*, c. 88, The Referendum (Grain Marketing) Act, 1931.

CHAPTER THIRTEEN

A Résumé of Experience with Contract Pools

CANADIAN WHEAT was marketed under the voluntary contract pooling system throughout the eight crop years, 1923–4 to 1930–1 inclusive. An interprovincial conference of grain growers held in Regina in July, 1923, had decided that pool organization, if and when it was accomplished, should be on a provincial basis, with a uniform contract and with provision for a single interprovincial selling agency. Sapiro's evangelical appearance in Alberta and Saskatchewan in August created tremendous enthusiasm for the contract pool plan and established among westerners, for the time being at least, a unity of purpose seldom equalled in the history of agrarianism. The contract sign-up campaign instituted in the individual provinces immediately following Sapiro's visit envisaged a minimum commitment of 50 per cent of the acreage of each province as a prerequisite to the introduction of the pooling system. In Alberta the goal was so near to attainment within a short time after the opening of the campaign that a pool was organized and arrangements were completed for the acceptance of deliveries on contract for a substantial part of the 1923 crop. The Alberta Co-operative Wheat Producers' Association, Limited, began receiving pooled wheat on October 29, 1923, and during the remainder of the crop year handled approximately 34 million bushels, or one-quarter of the wheat crop in the province. In Saskatchewan and Manitoba it was found impossible to secure enough signatures to complete arrangements for pooling in the fall of 1923. Organization proceeded throughout the crop year and the pools in the latter provinces were in position to handle grain at the commencement of the 1924 crop season.

The question of elevator facilities troubled membership and management alike during the early years. Efforts to secure satisfactory handling agreements with the co-operative elevator companies other than on a temporary basis were unsuccessful. Each of the provincial pools organized a subsidiary elevator company and embarked on an individual programme to construct and acquire elevators. Since a substantial number of the members of the pools were also members of the co-operative elevator companies, serious attention was given to proposals that the pools should purchase the facilities of the elevator companies.

Negotiations with the United Grain Growers came to naught but the Saskatchewan pool reached an agreement with the Saskatchewan Co-operative Elevator Company for the purchase of the latter's facilities at an arbitrated price of approximately $11 million. The transfer was effected on August 1, 1926, and the Saskatchewan pool was thus provided at a stroke with an addition to its elevator system amounting to 451 country houses and four terminals. Meanwhile the Saskatchewan pool and the other provincial pools as well had been actively engaged in constructing new elevators and they continued to do so until the end of the pooling period in 1931. By that time these organizations controlled approximately one-third of all the country elevators in the prairie provinces. The expansion of the provincial systems of country elevators is shown in Table VIII.

TABLE VIII

COUNTRY ELEVATORS OWNED AND OPERATED BY THE
PROVINCIAL POOLS, 1925–50

Crop Year	Man.	Sask.	Alta.	Total
1925–26	8	89	3	100
1926–27	30	586	42	658
1927–28	58	727	162	947
1928–29	143	970	314	1,427
1929–30	155	1,046	439	1,640
1930–31	153	1,060	438	1,651
1935–36	153	1,085	433	1,671
1939–40	155	1,093	424	1,672
1949–50	221	1,153	488	1,862

SOURCES: Manitoba Pool Elevators, *Annual Report*, 1951, p. 38; Saskatchewan Co-operative Wheat Producers Limited, and Alberta Wheat Pool, annual reports, *passim*.

The simultaneous expansion which took place in the terminal field involved construction, purchase, and lease. Before the end of the pooling period the pools controlled eleven terminals on the Great Lakes and the Pacific Coast with a combined capacity of over 36 million bushels, or more than one-third of the terminal capacity of the country at that time.

The contract which the individual grower made with the pool ran for five years, and called for the following deductions from the proceeds of his deliveries: (a) non-recurring deductions of $1.00 for one share of capital stock and $2.00 for organization expenses; (b) 2 cents per bushel of wheat (with varying amounts for coarse grains) for elevator reserve; and (c) 1 per cent of the gross proceeds of sales for commercial reserves. To finance the crop movement of members, or, in other words, to make provision for the initial payment, the pools effected an agreement with the commercial banks whereby the banks made advances to the pools on the security of warehouse receipts. The requirement was that the market value of the collateral should at all times exceed the banks' advances by a minimum of 15 per cent. This provision came to have critical significance for the pools when wheat prices collapsed in 1930.

Another pledge which the individual grower made in his contract was that he would deliver annually to the pool all the wheat which he disposed of commercially.[1] The first contract period covered the crop years 1923-4 to 1927-8. The second period was terminated by the waiving of all membership contracts in 1931, under the stress of demoralization in world wheat markets and following the failure of the campaign in Saskatchewan for compulsory pools. Membership figures which at their maximum were similar in the two contract periods exceeded 140,000 in the prairie provinces.[2]

The wheat pools marketed upwards of one billion two hundred million bushels of wheat in the eight years during which the growers' delivery contracts were in force—slightly more than one-half of the total wheat marketed in the prairie provinces throughout the same period of time. The annual figures are indicated in Table IX. This Table also indicates the pooled cash return per bushel of No. 1 Northern wheat at Fort William, year by year. The figures quoted as total payment represent initial, interim, and final payments for the respective years.

A great deal of controversy took place in the later pool period over whether the annual returns secured by pool members were above or below open market prices.[3] The controversy was futile for a number of reasons. There was, in the first place, a wide range of open market

[1]Each of the provincial pools organized a voluntary contract pool for coarse grains (oats, barley, rye) and flax, in addition to the pool for wheat. Coarse grains pools were in operation in Saskatchewan and Manitoba in time to handle the crop of 1925; in Alberta one became operative in 1929.

[2]The Ontario Grain Pool was formed in 1927 and had 13,200 contract members. Total Canadian memberships in the late 1940's exceeded 208,000. These, however, were memberships which carried no commitment concerning patronage.

[3]There was, for example, the incident relating to Appendix XII of the *Stamp Report* of 1931. Appendix XII, or Chart 10 as it was alternatively called, was offered as evidence before the Stamp Commission by a witness for the grain trade. The chart purported to show that open market prices had substantially exceeded pool prices for wheat throughout the years of pool activity. It was withdrawn from the *Stamp Report* following strenuous protests that it was irrelevant and after certain of the commissioners had denied knowledge of its inclusion. A facsimile of the chart may be found in the contemporary daily press. See, for example, *Leader-Post*, Regina, June 20, 1931, p. 17.

TABLE IX

WHEAT POOL DELIVERIES AND PAYMENTS, 1923-4 TO 1930-1

Crop year	Deliveries to the pools (millions of bushels)	Pool deliveries as percentage of total western deliveries	Pool price per bushel (basis: No. 1 Northern, Fort William)	
			Initial payment	Total payment*
1923–24	34.1	26.0	$0.75	$1.01
1924–25	81.7	37.3	1.00	1.66
1925–26	187.4	52.2	1.00	1.45
1926–27	180.0	53.1	1.00	1.42
1927–28	209.9	51.1	1.00	1.42¼
1928–29	243.9	51.3	0.85	1.18½
1929–30	121.7	52.0	1.00	1.00
1930–31	126.7	41.2	0.70	
			0.60	0.67
	1,185.4		0.50	
			0.50	

*Including reserve fund deductions.
SOURCE: Adapted from data in *Tides in the West*, published by Alberta Wheat Pool, Dec., 1936, App., p. iv.

prices from which to pick "the" open market price, and no one of them was obviously the only one to use. There was, in the second place, a great variety of ways of striking a proper average in order to secure a single annual figure comparable with that ready to hand in wheat pool calculations. Finally, there was no way of knowing what the open market prices would have been had the pools not been in operation. The question whether a member of the wheat pool was better off financially than his neighbour who was not a member, even if it could have been answered, was not the same as the question whether he was better or worse off than he would have been if no pool had been in existence.

A financial crisis of major proportions rather than any decision of policy brought an end to the contract pooling of Canadian wheat. Regardless of internal friction, the Canadian pools weathered the first five-year contract period. A new sign-up campaign was successfully conducted with new or renewal contracts roughly equal to the old. Five crops had been satisfactorily disposed of and the Central Selling Agency entered the new contract period with no carryover from previous crops.

Special difficulties arose with the marketing of the crop of 1928, but the effects of these difficulties did not become serious until the problems faced by the pools a year or more later were added to them. The Canadian wheat crop of 1928 was of record size, surpassed by only two other crops to the present time, those of 1952 and 1953. It was, however, comparatively poor in quality and was brought to world markets which were already heavily stocked from other sources. The prospects, although imperfectly known, were clearly bearish in the early summer of 1928 when the pools faced the necessity of determining the magnitude of the initial payment for the new crop. They had set an initial payment of $1.00 per bushel, basis No. 1 Northern, Fort William, for each of the crop years 1924–5 to 1927–8, and their margin with the lending banks had rarely been threatened and had never been impaired throughout that period. The bearishness of the circumstances in 1928 led them to decide on an initial payment of 85 cents per bushel.

Wheat deliveries to the 1928–9 pool totalled 244 million bushels and despite record exports the pools ended the crop year with a carry-over of 48.4 million bushels, mostly of the lower grades. Interim payments had meanwhile brought the total advance to farmer members to $1.18½ per bushel, basis No. 1 Northern, Fort William, with the other grades in proportion.

Despite the carryover from the 1928 crop, the general market situation was far from bearish in the summer of 1929. World wheat production appeared to be substantially reduced and Canadian prospects indicated a sharp reduction in output from that of the previous year. The Winnipeg price for No. 1 Northern wheat (in store at Fort William) had fluctuated roughly between $1.15 and $1.30 per bushel throughout the latter half of 1928 and the early months of 1929. A sudden speculative flurry in midsummer, 1929, carried the price to peaks of $1.78¾ in July and $1.73⅝ early in August. On July 11, 1929, the pools authorized an initial payment of $1.00 for No. 1 Northern for the new crop year, an increase of 15 cents per bushel over the initial payment of the previous year and, in effect, a restoration of the level of the four preceding years. That, in conjunction with the margin requirements of the lending banks, meant, in general terms, that prices had to remain at levels above $1.15 per bushel, No. 1 Northern, Fort William, if the credit standing of the pools was to be kept intact. The collateral margin of the pools was not impaired until early in 1930 but the banks and the pools were understandably concerned at the weakness of produce markets and the persistence of price declines long before that time. In mid-November, 1929, on the morrow of the collapse of the New York stock market, pool officials discussed the financial position of their organization with representatives of the lending banks. McPhail noted in his diary on the fourteenth of the month that "the banks have been rather uneasy and have been a little troublesome."[4] A further meeting,

[4]H. A. Innis, ed., *The Diary of Alexander James McPhail* (Toronto, 1940), p. 209. Hereafter cited as *McPhail Diary*.

in mid-January, 1930, found the bankers affable and reassured. Soon after that, however, No. 1 cash wheat fell to $1.20¾ in Winnipeg and on January 29 the chairman of the lending banks' committee (Mr. Cork) gave pool officials to understand that any impairment of the collateral margin would lead the banks to demand immediate liquidation of stocks. The pools immediately sought and secured the agreement of their respective provincial governments to guarantee the banks against loss on their advances. And none too soon, for the low price in February was $1.07⅞, and in March, $1.00⅞. In June the price dropped below $1.00 a bushel for the first time in fifteen years. The provincial guarantee permitted the Central Selling Agency to continue with full autonomy throughout the 1929–30 crop year and, despite a good deal of annoyance associated with the arrangement of terms, the pools were further permitted to embark on the receipt and disposal of the 1930 crop as it came to market.

Conditions went from bad to worse during the early months of the 1930–1 crop year. Each month after midsummer, 1930, the price of wheat fell lower. The average price of No. 1 Northern wheat in Winnipeg for the successive months from August to December was 92½, 78⅛, 72½, 64⅜, and 55¾ cents. The initial payment which was set tentatively, effective July 16, at 70 cents per bushel, No. 1 Northern, was reduced to 60 cents on August 26, to 55 cents on October 15, and to 50 cents on November 11.[5]

The collapse of market values and the difficulty of selling wheat created a financial crisis for the pools which deprived the Central Selling Agency of autonomy and brought contract pooling of wheat and coarse grains to an end in Canada. The transition was effected during the two crop years 1929–30 and 1930–1, the first years of the onset of the world's worst depression. During these years the distress of producers in the prairie provinces was acute but differed only in degree from that of producers in the other sections of the Canadian economy.

Prime Minister Bennett was approached by the western premiers on August 7, 1930, the day on which his government took office. A series of conferences which he held with the provincial premiers, the bankers, and pool officials apparently did not commit his government to guarantee the banks' advances to the pools but cleared the way for a tri-party

[5]See Canadian Co-operative Wheat Producers Limited, *Directors' Report*, 1930–1, p. 4, reprinted in Alberta Wheat Pool, *Annual Report*, 1930–1. The pools ended the 1929–30 crop year with a carryover of 67 million bushels, and part of the problem of arranging for the financing of the 1930 crop concerned the question of apportionment of subsequent sales between the old and the new crop holdings. See *ibid.*; also *McPhail Diary*, 216 ff.

agreement among the governments of the prairie provinces, the banks, and the pools regarding apportionment of sales in the new pool period.

The Imperial Conference took Canadian political and pool leaders to London for two months following the middle of September. McPhail found on his return that the pools were on the verge of collapse, with the banks reluctantly and tardily advancing the funds necessary to meet the Central Selling Agency's position in the Winnipeg clearing house from day to day.[6] Federal ministers refused formal commitment in regard to the pools pending the return of Mr. Bennett from the Conference. Eventually the federal government agreed to guarantee the pools' advances provided the Central Selling Agency was placed under the control and management of some person acceptable to the banks and the government. John I. McFarland, formerly president of the Alberta Pacific Grain Company and an outstanding figure in the private grain trade, was suggested. On being assured that his appointment would be acceptable to all parties[7] McPhail asked him to become general manager of the Central Selling Agency. McFarland stated as a condition of his appointment that he would expect to close the London office as a move to "get more friendly with the grain trade and eliminate antagonisms."[8] McPhail told him that "as a result of him being unanimously approved by the banks and the Premiers [the pools] were largely under his control for the time being."[9]

[6]On November 14, 1930, McPhail wrote: "Market fell today. We were two hours late in getting a cheque for $211,000 over to the clearing house as a result of the reluctance of the banks to advance the money. As a result of the delay and the fear it inspired of our position, Frank Fowler [manager of the clearing house] notified us he would make a call Saturday morning for another 5 cents a bushel on 7,000,000 long. Premiers met banks in afternoon and again at 9 p.m. Everything is tottering." And on the fifteenth: "All day at the office. We had quite a time getting a cheque for $300,000 to put in the clearing house. The banks last night agreed to put up margins until Monday. Spears told me this morning that the Royal Bank had backed out of putting up their share, and the Commerce and Montreal put it up. We met the banks at 12. They had word from Toronto not to put up any more margins and we had to have $200,000 more up at 1.15. They finally agreed to stand by their agreement of last night. Otherwise the pool would be bust." *McPhail Diary*, pp. 227–8.

[7]Including the premiers of the prairie provinces, the banks, and the Prime Minister. Speaking of the matter of selecting a new general manager for the Central Selling Agency, McPhail commented: "The bankers are not particularly interested in anything pool officials have to say. They are only interested just now in anything that will help to get the federal Government to protect them against loss. Hence the reason they are so set on McFarland becoming general manager. They know he is one of, if not the most intimate friend of, Bennett's and that if he can be got into the pool it will probably do more than anything else to get the Dominion Government to do something." *Ibid.*, p. 230.

[8]*Ibid.*, p. 231.

[9]*Ibid.*

By the end of November, 1930, McFarland was installed as general manager of the Central Selling Agency.[10] This accomplished, the federal government guaranteed the banks' advances to the agency for the 1930 crop. McFarland's first official act was to close the London office of the Central Selling Agency, and very soon he withdrew the other overseas agencies which had constituted the instrument for implementing the pools' policy of direct marketing. The board of directors of the Central Selling Agency continued to meet and for some time attempted to maintain the fiction that the general manager, if not under the control of the board, was nevertheless concerned to hear the expression of their opinions on marketing policy. McFarland quickly dispelled this illusion, making it clear that the condition of his gratuitous services was a free hand in disposing of the pools' stocks of wheat, subject only to the dictates of the federal government as guarantor of financial commitments.[11]

The pools entered the 1930–1 crop year with a substantial carryover from the 1929 crop together with a small carryover from the 1928 crop. Early in 1930 it became apparent that, barring a miraculous recovery of the wheat market, pool members had been overpaid on the basis of the initial payment of $1.00 per bushel, No. 1 Northern. The amount of the overpayment could not, of course, be determined until the last of the 1929 crop, as well as the remnant of the 1928 crop, had been disposed of. Meanwhile the funds involved in the overpayment constituted an advance from the banks to the pools, protected after the early months of 1930 by the guarantee of the prairie provincial governments, and from November, 1930, by the guarantee of the federal government as far as the crop of 1930 was concerned. The area of provincial liability was thereby delimited and determination of the dollar amount merely awaited the final disposal of the grain or futures remaining from the 1928 and 1929 crops.

The small balance of the pool holdings of the 1928 crop was cleared out in May, 1931, and the much more substantial carryover from the

[10]He served without salary, that being one of his own conditions for accepting the position. Canadian Co-operative Wheat Producers Limited, *Directors' Report, 1930–1*, p. 5.

[11]On March 11, 1931, McPhail wrote: "Central Board meeting all day. We had a visit from McFarland. He is like a fish out of water in the Board room. He has no sympathy with, or understanding of, the aims of the farmers' movement." And on the twelfth: "Board meeting all day. Adjourned at 5 P.M. to go to Regina for a meeting of the three Boards tomorrow. When Wesson asked McFarland if he was coming to Regina, McFarland said he thought the boys would be better back on the farm." On May 14: "McFarland met the Board but we did not get any information from him if he has any to give. He is very unsatisfactory. Had no regard nor does he care what the Board thinks." *McPhail Diary*, p. 236.

1929 crop was finally liquidated in June of the same year. The net overpayment of the 1929 crop after deducting the proceeds from the remainder of the 1928 crop of wheat and coarse grains (approximately $2,750,000) amounted to more than $22 million. The provincial allocation and the total sum as at August 31, 1931, were as follows:[12]

Manitoba	$ 3,364,722
Saskatchewan	13,265,054
Alberta	5,520,162
TOTAL	$22,149,938

As a result of negotiations among the three pools, the seven creditor banks, and the three provincial governments, an agreement was reached whereby the governments issued 4½ per cent bonds (at 2 per cent discount) to the banks to cover their respective share of the liability. The provincial pools in turn issued 5 per cent, 20-year amortization bonds to their respective governments in corresponding amounts. These bonds were secured by the elevator assets of the provincial pools. In Saskatchewan and Alberta the value of the elevator assets was well above the liability involved in the pledge of security to the governments. The Manitoba pool was declared bankrupt in November, 1932, and the government agreed to absorb $1,400,000 of the overpayment liability to the banks and to accept security for the balance in redeemable stock of the reorganized Manitoba Pool Elevators.[13]

It may reasonably be argued that the moneys involved in the transactions discussed above did not represent *losses* sustained by the pools.

[12]Canadian Co-operative Wheat Producers Limited, *Directors' Report*, 1930–1, Statement 1, in Saskatchewan Co-operative Wheat Producers Limited, *Seventh Annual Report*, 1931, pp. 46–7. The settlement date was ultimately set at September 30, 1931, with the amounts adjusted for interest as of that date. The total sum adjusted to the latter date was $22,217,302. See Canadian Co-operative Wheat Producers Limited, *Directors' Report*, 1931–2, in Saskatchewan Co-operative Wheat Producers Limited, *Eighth Annual Report*, 1932, p. 36. The overpayment on the 1929 crop amounted to approximately 18 cents per bushel, with an excess of that amount on certain grades.

[13]These arrangements were made in the Manitoba "Four Party Agreement" of August, 1931, the four parties being presumably the provincial government, the banks, the Manitoba Wheat Pool (Manitoba Co-operative Wheat Producers Limited), and Manitoba Pool Elevators Limited. Referring to the bankruptcy of the provincial pool, the directors of the Manitoba Pool Elevators said in part: "Rumors have been widely circulated to the effect that Pool Elevators also were involved. These rumors have been officially contradicted in a statement by the President, and your Board desires to repeat here most emphatically that the action has reference only to the organization officially known as 'Manitoba Wheat Pool,' and that it has no connection whatsoever with the operations of Manitoba Pool Elevators Limited." See Manitoba Pool Elevators Limited, *Reports for the Financial Year ending July 31, 1932*, p. 5.

The co-operative agency for the handling of pooled wheat—the Central Selling Agency—had, through mistaken judgment, advanced to the growers a greater sum per bushel than was realized on sales. These sales of necessity took place over the months following the establishment and payment of the initial payment and were made on the most disordered and demoralized grain markets in modern history. The size of the initial payment, incidentally, had at least the tacit and pragmatic sanction of the lending banks in that they extended credit to the pools sufficient to make the initial payment possible. Nor would any one but the chronic pessimist have been sceptical about the feasibility of an initial payment of $1.00 per bushel in midsummer of 1929 when wheat prices were approximately 75 cents per bushel above that figure. The directorates of the wheat pools, democratically elected, faced a difficult task in establishing an initial payment under the best of circumstances. Growers who were members of the pools naturally desired as high an initial payment as possible because of the ever-present pressure for funds. For the directors to establish too high a payment would, nevertheless, be dangerous and inexcusable. To set too low an initial payment, as might be suggested by ultra-conservative counsel, would leave the members dissatisfied, would promote bootleg infraction of contracts, and would be "politically" impossible. Pool officials could not, in the summer of 1929, foresee the economic future. That inability was shared in equal measure by members of the private grain trade, by Canadian business men generally, and by the average citizen.

The federal guarantee for the 1930–1 crop year provided no indication of how the next crop, that of 1931, would or should be handled. The directorate of the Central Selling Agency considered it highly desirable to continue the pooling system if at all possible. McFarland scoffed at the idea of the banks being willing to finance disposal of another crop by the pools.[14] Differences of opinion developed within the central board, some members urging that the whole problem be left in the lap of McFarland as the representative of the federal government. Late in March, 1931, the Alberta board gave notice of withdrawal from the Central Selling Agency.[15] A month later (April 27, 1931) the compulsory pool campaign in Saskatchewan was liquidated by the decision of the Saskatchewan Court of Appeal, which ruled the Saskatchewan Grain Marketing Act *ultra vires*. Pool officials turned to the federal government with a request for the restoration of a national wheat board. Thus began anew the wheat board movement which

14*McPhail Diary*, p. 236.
15*Ibid.*, p. 238.

eventuated in the wheat board legislation of 1935. Meanwhile the pools abandoned the contract pooling of grain. In June, 1931, they terminated the delivery contracts of their members. Each of the provincial organizations established a voluntary pool and continued to offer marketing service on this basis for the four years preceding the establishment of the Wheat Board in 1935. However, only a small quantity of wheat, approximately 20 million bushels in all, was delivered to the voluntary pools.

McFarland had been appointed general manager of the Central Selling Agency in order to secure federal guarantees which would permit disposal of pool wheat, without chaotic liquidation, until the end of the 1930–1 crop year. His activities were financed under guarantee of the federal treasury, he was responsible to the federal government, and whatever reports he made were made in confidence to the federal ministry. The directorate of the Central Selling Agency had no information as to what McFarland, as general manager of the agency, was doing in regard to the sale of wheat. McFarland, in turn, claimed to be completely uninformed as to the terms of the agreement between the federal government and the banks concerning federal guarantee of the banks' advances for the current crop.[16] His sales activities were early transformed into stabilization activities[17] which continued for four years. In 1935, when a national wheat board was restored after a lapse of fifteen years, he was still the wheat marketing agent for the federal government and was an obvious choice as chairman of the new board. These matters, however, are appropriate for consideration in the next chapter rather than the present. It is only necessary to mention them here in order to indicate the circumstances under which contract pooling ended in Canada and gave way to government intervention.

The loss of autonomy by the Central Selling Agency and the liquidation of delivery contracts by the provincial pools at the end of the crop year 1930–1 left the pools in the position of three farmer-owned elevator systems with extensive facilities at both the local and terminal levels. Structurally, grain growers' co-operation was back where it had been a decade and a half before. There were now four co-operative elevator systems—including the United Grain Growers Limited—where at the end of the First World War there had been two. The four systems were collectively very large, several times as extensive as the farmers' elevator

16Ibid., p. 235.
17As early as the latter part of January, 1931, McFarland informed McPhail "again" that he had authority from the Prime Minister to keep the price of wheat from falling below 50 cents per bushel. See ibid. Stabilization purchases began in July, 1931.

systems had been at the end of the war. The facilities of the Saskatchewan pool were as extensive as those of the other three farmers' companies combined. The four systems comprised over two thousand country elevators and a dozen terminals.

Throughout the twenty-year period following the termination of contract pooling in 1931, the three provincial elevator systems prospered,
expanded, and, to a certain extent, diversified their activities. Each one
liquidated in full its financial obligation to its respective provincial
government in connection with the financing of the 1929 overpayment.
The total payments by the three pool elevator systems on this account
were in excess of $34 million, including $12.3 million in interest.[18] The
Alberta pool made its final payment on June 1, 1947, the Saskatchewan
pool in September, 1949, and the Manitoba organization a month
later.[19] By the end of the 1950–1 crop year the three pool elevator systems had declared patronage dividends to a total of approximately $50
million.

Since 1931, the commercial activities of the Alberta and Manitoba
pool organizations have centred almost exclusively on the operation of
their elevator systems; the Saskatchewan pool has diversified its interests
to a certain extent, although the operation of the elevator system has
continued to be its predominant function. In May, 1944, the Saskatchewan livestock pool (Saskatchewan Co-operative Livestock Producers
Limited) was amalgamated with the Saskatchewan wheat pool. The
wheat pool changed its own name from Saskatchewan Co-operative
Wheat Producers Limited to Saskatchewan Co-operative Producers
Limited, and it has since operated the livestock pool under the latter's
name, Saskatchewan Co-operative Livestock Producers Limited. The
livestock pool thus forms a subsidiary of the parent company on the
same basis as the elevator system which is a subsidiary called officially
Saskatchewan Pool Elevators Limited. Pursuing a general interest in
industrial development and in expanding the industrial uses of agricultural products, the Saskatchewan pool opened a vegetable oil extraction
plant in November, 1946, and a flour mill with a capacity of 1,000
barrels daily in February, 1949. These plants are operated by Saskatchewan Pool Elevators Limited, the main operating subsidiary of the parent
pool.

[18]Canadian Co-operative Wheat Producers Limited, *Directors' Report*, 1948–9,
in Saskatchewan Co-operative Producers Limited, *Twenty-Fifth Annual Report*,
1949, p. 39.
[19]*Ibid.*

The Central Selling Agency, its commercial functions gone, has carried on as a liaison organization, collecting and disseminating information and providing a focus for the expression of the views of pool membership on grain marketing policy, not only at the national but also at the international level.[20] Starting in 1931 the central directorate spearheaded the drive for the re-establishment of a national wheat board which culminated in the establishment of such a board in 1935. Thereafter it urged the suspension of futures trading in Winnipeg and pressed to have the wheat board designated the exclusive agency for the marketing of Canadian wheat and of coarse grains and flax as well. Acting directly, as well as through the Canadian Federation of Agriculture— of which organization the pools are members and strong supporters—the central board made repeated representations to the federal government throughout the war years concerning production and marketing policy. The board continued to concern itself in matters lying far beyond the national field as it had in the years of contract pooling. At international conferences of agricultural producers, its officers have frequently been spokesmen for the organized wheat growers, whose deep-seated aversion to open market trading led, in the 1920's, to pooling and direct overseas marketing. These gone, pool leaders revived the alternative ideas of state trading and multilateral agreement among the governmental agencies of a maximum proportion of the producing and consuming countries.[21] They were active participants in conferences leading to the unsuccessful international wheat agreement of 1933 and to the wartime agreement of 1942 which was not put into effect. After the war they gave their full support to the efforts which led to the International Wheat Agreement of 1949 and to its renewal in 1953.

[20]In its report for 1951–2 the Central Selling Agency says in part: "Your Central Board is the instrument through which the opinions, needs and decisions of the producers who make up our three Canadian Wheat Pools, are brought into final focus—for united action in their common interests. A salient issue facing the farmers of our Canadian West is the International Wheat Agreement. The decision of farmers—expressed at hundreds of Pool meetings held during the past year in neighborhoods all over the West—was that the Agreement should be renewed. . . . This decision was carried, through your Central Board, into the councils of our Canadian nation; and, finally to the International Wheat Council." Canadian Co-operative Wheat Producers Limited, *Directors' Report*, 1951–2, in Saskatchewan Co-operative Producers Limited, *Twenty-Eighth Annual Report*, 1952, p. 76.

[21]Commenting on the negotiation of the International Wheat Agreement of 1942, the central board traced the support of the pools for the international organization of wheat marketing back to 1927. See Canadian Co-operative Wheat Producers Limited, *Directors' Report*, 1941–2, in Saskatchewan Co-operative Wheat Producers Limited, *Eighteenth Annual Report*, 1942, pp. 69–70.

CHAPTER FOURTEEN

From Open Market to Compulsory Board

THE DOMINION GOVERNMENT became heavily involved in the marketing of wheat with the onset of the depression in the early 1930's and has continued so to the present time. There were important questions of principle implicit in the Dominion's original intervention but the occasion was of such impelling necessity that there was little opportunity for their consideration.

McFarland's activities as head of the Central Selling Agency and on government account continued from November, 1930, until the autumn of 1935, when the second Canadian Wheat Board was established.[1] He was then appointed chairman of the Wheat Board and acted in that capacity until December 2, 1935, when he and the other members of the Board were retired by the newly elected Liberal government. Mr. McFarland's activities came to be known as stabilization activities and the period from 1930 to 1935 is commonly identified as the period during which the Dominion government engaged in stabilization operations in the Canadian wheat market. This terminology is correct enough in retrospect but it conceals the transformation in purpose and in method which took place inconspicuously in the period between the practical liquidation of the Central Selling Agency in December, 1930, and the establishment of the Wheat Board in August, 1935.

McFarland's appointment to the agency was made and the pools' overdraft was originally guaranteed by the federal government with no intention on its part to commence protracted stabilization operations. The government acted initially to avert the catastrophe which would inevitably accompany the dumping of the pools' holdings of wheat from the crops of 1928, 1929, and 1930 on a market which was already seriously demoralized. There was no way of foretelling all the results which would follow an unbridled sale of Canadian wheat stocks. It would certainly ensure the collapse of the pools as pooling agencies, but that was of little concern except to their officials and members. The more significant possibility in the imminent collapse of the wheat pools was that the closely knit banking and financial structure of the Dominion

[1]The Canadian Wheat Board Act, *Statutes of Canada*, 25–26 Geo. V, c. 53, was assented to July 5, 1935. Board membership was announced August 14.

might be imperilled. This risk, whether regarded as substantial or slight, was one to be avoided at any cost.

The pools entered the 1930–1 crop year with a carryover of 67 million bushels of wheat and took delivery of 127 million bushels from the 1930 crop. The total, almost 200 million bushels, while by no means all in the hands of the Central Selling Agency at any one time, indicates the magnitude of the task which confronted the pools in the autumn of 1930 and which was assumed by McFarland in November of that year.[2] McFarland's first view was that his assignment was to sell wheat and to clear out the pools' holdings rather than to withhold their stocks in order to relieve the market. He shared the view which was prevalent outside pool circles that the pools' practice of selling directly through their own overseas agencies had fostered and consolidated an effective degree of sales resistance to Canadian wheat. If this view were correct, the withdrawal of overseas agencies and the restoration of the trade to other channels might well make possible the early liquidation of Canadian wheat surpluses. Reasoning in this fashion, McFarland closed the pools' overseas agencies in December, 1930, immediately following his appointment as general manager of the Central Selling Agency, and exerted every effort to move the agency's wheat into consumption channels. In these efforts he met with some success. Canadian exports, pool and non-pool combined, totalled 258 million bushels or 40 per cent of the world trade in wheat during the crop year 1930–1.

McFarland's desire to sell wheat did not lead him to disregard the matter of price or to ignore the disastrous implications of the decline which persisted throughout the months following his appointment to the Central Selling Agency. At the end of January, 1931, he commented to McPhail that he had authority from the Prime Minister to keep the wheat market from going below 50 cents a bushel. He denied, however, that this authority was of decisive importance in determining his selling policy. He said, "I am not going to sell at these levels. If the Government or the banks or the pool want any more wheat sold now at these prices they can put in someone else. I won't do it."[3]

The world wheat market, as part of the total world economic situation, was, in any case, much too disturbed by 1931 for the withdrawal of overseas agencies to accomplish the miracle necessary for its recovery. In addition to the competition of supplies from Argentina and Australia,

[2]Cash and futures wheat holdings of the Central Selling Agency in November, 1930, were approximately 37 million bushels. See Table X below.
[3]As reported by McPhail in H. A. Innis, ed., *The Diary of Alexander James McPhail* (Toronto, 1940), p. 235.

supplies which were made exceptionally bearish by exchange depreciation in these countries, there was competition from heavy offerings of
Russian wheat, which appeared on the world market for the first time
since 1914. American stabilization supplies, which had been accumu-
lating in the hands of the Federal Farm Board since 1929, overhung the
market. Despite the heavy exports of the crop year 1930–1, the Canadian carryover on July 31, 1931, reached the record total of 140 million
bushels, and wheat prices had declined to little more than 50 cents per
bushel, basis No. 1 Northern, Fort William. McFarland's agency had
converted most of its unsold holdings into futures and ended the crop
year with approximately 75 million bushels of wheat or wheat futures
in its account.

Under these circumstances it became increasingly clear that the
liquidation of pool holdings was but a part of a continuing problem and
by no means its most crucial element. With terminal wheat prices sagging
toward 50 cents a bushel, the provincial pools released members from
their contracts in June, 1931, and instituted voluntary pools for succeeding years. This left the disposition of Canadian wheat, with the exception
of the amount still held by McFarland's agency, exclusively in the hands
of the private trade on the pre-pool open market basis. McFarland's
original assignment had contemplated only the liquidation of pool wheat
and, as of July 31, 1931, this would have required only the gradual sale
of the 70-odd million bushels of wheat and wheat futures which remained
in hand. If the original design had been carried to conclusion, Mc-
Farland's activities as grain marketing agent for the Dominion government would have ended much sooner than they did and would not have
come to be characterized by the name of stabilization operations.[4]

McFarland's activities on government account were extended and
transferred into positive price support operations in 1931. In June of
that year he was given authority to purchase wheat or wheat futures
under government guarantee for the purpose of providing some degree
of strength to the current market. There was apparent a complete lack of
speculative willingness to absorb the hedging sale of futures which accompanied the disposal of Canadian wheat, and over the succeeding
four years this condition persisted or at least frequently recurred. Mc-
Farland's purchases for stabilization purposes were first made in July,
1931, and they continued in varying proportions until the establishment
of the Wheat Board in 1935. Pressure on world wheat prices was particularly heavy during the latter part of 1932 and despite substantial

[4]Stabilization may, of course, be attempted by the negative action of with-
holding supplies as well as by the positive action of purchasing additional supplies
as they come on the market.

purchases of futures in an attempt to strengthen the market, the Winnipeg price of No. 1 Northern fell below 40 cents per bushel on December 16 and 20, 1932, a price which had no parallel in its world counterpart throughout the preceding four hundred years.

Stabilization holdings at times between 1931 and 1935 represented practically the entire stocks of Canadian wheat in marketable position. The maximum accumulation of McFarland's agency was 235 million bushels in October, 1934, and this involved a bank overdraft under government guarantee totalling $80 million. Table X indicates the quantities of wheat held on stabilization account at various significant dates.

TABLE X

WHEAT AND WHEAT FUTURES HELD BY CANADIAN CO-OPERATIVE WHEAT PRODUCERS ON SPECIFIC DATES THROUGHOUT THE PERIOD OF STABILIZATION OPERATIONS, 1930–5

Date	Bushels
Nov., 1930	36,935,000
July 31, 1913	75,164,000
July 31, 1932	99,978,000
July 31, 1933	149,672,000
July 31, 1934	176,237,000
July 31, 1935	213,688,000
Dec. 2, 1935	205,187,000

SOURCE: Table adapted from data in T. W. Grindley *et al.*, *The Canadian Wheat Board, 1935–46* (Ottawa: King's Printer, 1947), p. 5 (reprinted from *Canada Year Book*, 1939 and 1947 editions).

Dominion stabilization operations which emerged from expediency and almost without design in 1931 came to an end only after the establishment of the second Canadian Wheat Board in 1935. The Central Selling Agency turned 205 million bushels of wheat and wheat futures over to the board on December 2, 1935, and active disposal of the stabilization stocks commenced. It was not until the end of the crop year 1937–8, however, that the board was able to report the final liquidation of the "old wheat" which it had acquired (largely in the form of futures) from the stabilization account.[5] When the final balance was struck, it indicated that the total protracted stabilization operation had been closed out at a net profit of approximately $9 million after taking full account of carrying charges and other costs.

Stabilization operations constituted but one of several expedient attacks made by the Bennett government on the Canadian wheat prob-

[5]See *Report of the Canadian Wheat Board, Crop Year 1937–1938*, p. 2.

lem after 1930. In 1931 the government paid the growers a bonus of 5 cents per bushel on the crop produced that year. This payment appears almost ludicrously small in relation to the price of wheat and the value of money today. It appears in a more favourable perspective when it is noted that, apart from the bonus, the estimated farm price for wheat in the calendar years 1931 and 1932 was 38 and 35 cents respectively. Total bonus payments on the 1931 crop amounted to $12,720,000. Apart from any question of adequacy or inadequacy, however, the basic flaw in a bushel bonus under any circumstances—including, of course, price support measures—is that the benefits go to persons who have a crop and in direct proportion to the size of their crop, whereas persons who suffer a crop failure gain nothing.

Mr. Bennett's government placed great faith in the efficacy of fiscal adjustment as a means of attack on the economic disorder of the early 1930's. Increased domestic tariffs adapted to a revitalized and generalized system of Empire preference should serve—so it was argued with a peculiar economic sophistry—to restore a desirable amount and direction of Canadian trade. The faith of the Conservatives in these measures was so great that it is not surprising to note their attempt to fit the trade in wheat into the over-all pattern of Empire preference. British political leaders proposed an Empire milling quota but Mr. Bennett urged and secured at the Ottawa Conference of 1932 an Empire preference of 2 shillings a quarter, or approximately 6 cents a bushel. The preference, like the bushel bonus of the year before, was small and its effectiveness would have been negligible on that account if on no other, but like the bushel bonus its most serious defect lay in the principle behind it. The British Empire and Commonwealth constituted a wheat surplus area in the 1930's as before and since. Under such circumstances no system of internal preference, however generous, could have anything but minor efficacy.[6] Western agricultural leaders clearly recognized this elementary principle and pool officials advised Mr. Bennett that they had no interest in either Empire quotas or Empire preference as a result. The preference, instituted in 1932, was nominally in force until its removal in January, 1939, but it contributed nothing toward the solution of the Canadian wheat marketing problem.

Considerably greater vision but no appreciably greater immediate results characterized another venture in the field of international negotia-

[6]Dr. MacGibbon, however, analyses the effectiveness of the imperial preference on wheat in relation to circumstances of seasonal supply and demand in the Atlantic region. He is not inclined to dismiss its influence as negligible. See *The Canadian Grain Trade, 1931–1951* (Toronto, 1952), pp. 21–5.

tion by the Bennett government in its attempt to restore the market for Canadian wheat. This was its sponsorship of the London Wheat Conference of August, 1933, and its support of the London Wheat Agreement which resulted from the conference. The conference was called by the Secretary General of the League of Nations on the request of the four principal wheat exporting countries, representatives of which had already held a preliminary conference in Geneva under the League's auspices. Mr. Bennett presided over the London meetings. The agreement, signed by representatives of twenty-two importing and exporting countries, was to apply to the two crop years 1933–4 and 1934–5 and involved the acceptance of export quotas but no pre-determination of price. The quotas for 1933–4 rested on an estimated world trade requirement of 560 million bushels, and for 1934–5 the plan was to reduce exports, in line with an elaborate formula, by 15 per cent from a vaguely defined norm.

The London Wheat Agreement contributed little if anything to the solution of the world wheat problem of the 1930's. It broke down under the failure of certain of the signatories, notably the Argentine, to abide by their quota restrictions on exports. A wheat advisory committee was set up in conjunction with the agreement, however, and on the outbreak of the Second World War this committee was actively engaged in an attempt to formulate a new agreement. The United States Department of Agriculture was an active sponsor of these efforts as it had been in the promotion of the 1933 agreement. Speaking in Washington on September 5, 1939, Secretary Wallace said: "Before the outbreak in Europe there was, I believe, substantial ground for hoping that an effective wheat agreement would be reached very soon."[7] Further action awaited the conclusion of the war and was, in fact, postponed for several years thereafter. The International Wheat Agreement of 1949 was, however, by no means a new venture nor was it negotiated without reference to the unsatisfactory experience with the London Wheat Agreement of 1933.

Western agricultural leaders were active supporters of the attempts made in the early 1930's to secure international co-operation in the marketing of wheat. They increasingly came to regard Mr. McFarland's stabilization operations as helpful in averting the worst of the impact of international deflation on the market for Canadian wheat. They were not impressed by any other of Mr. Bennett's efforts to solve the wheat problem. Year after year they urged the Dominion government to estab-

[7]See *Wheat Studies*, vol. XVI, no. 4, Dec., 1939, p. 140 n.

lish a national wheat marketing board as the only type of agency competent to meet the immediate and long-run requirements. In May, 1931, immediately after it became clear that the 100 per cent compulsory pool campaign in Saskatchewan had failed, McPhail, in a radio address, urged the Dominion government to establish a wheat board in preparation for the conference which was to take place in London later in the month. In June, he and the premiers of the three prairie provinces met Mr. Bennett to give the same advice. Repeatedly thereafter, western agricultural leaders reiterated their request and made it plain that nothing less than a national marketing board would satisfy them. They finally gained their objective in substantial measure with the establishment of the second Canadian Wheat Board under legislation assented to in July, 1935.

The duration and continuity of the grain growers' belief in the superior efficacy of a national marketing board must be emphasized. Statements made by western agricultural leaders substantiate much fragmentary evidence to the effect that the desire for a national board was never entirely submerged throughout the years when the wheat pools were active. Mr. R. S. Law of the United Grain Growers informed the Special Committee of the House of Commons which considered the wheat board bill in 1935 that, "To some extent some feeling in favour of a resumption of governmental activity similar to that of 1919 persisted even in the most successful time of pool operation."[8] Mr. Brouillette, then president of the Saskatchewan Wheat Pool, stated before the same committee: "Ever since the Canada Wheat Board of 1919–20 suspended operations there has been a strong desire on the part of the vast majority of western farmers for the re-establishment of a national Wheat Board."[9] In the annual report of the Canadian Co-operative Wheat Producers (the Central Selling Agency for the pools) for 1934–5 the directors voiced at length their gratification that a national wheat board had been re-established. They said in part:

Your Directors feel that the passing of this legislation [creating the new Board] is one of the most progressive steps taken by any Parliament in Canada to deal with a situation that affects the entire Dominion. It is a matter of very great satisfaction to the organized producers that the principle for which they have so long contended—that the wheat problem is a national problem—has received the approval of all parties in our Canadian Parliament. . . .

[8]See Special Committee on Bill 98, Canadian Grain Board Act, *Minutes of Proceedings and Evidence* (Ottawa, 1935), p. 83.
[9]*Ibid.*, p. 187.

Your Board, during the past four years, has persistently urged and supported legislation for the Canadian Grain Board to control the marketing of grain in the farmers' interests and the national well-being. . . .

The establishment of a National Wheat Board with complete control over the marketing of wheat has been urged by the farmers of Western Canada since 1919 and is in accord with the developments in other wheat countries. . . .[10]

In a radio address in November, 1935, F. W. Ransom, secretary of the Manitoba Pool Elevators, reported on the various "milestones" of achievement in the recent record of the organized grain growers. "The [latest] milestone," he said, "is called 'The Canadian Wheat Board,' and it tells a story of an effort on the part of the organized farmers covering a period of seventeen years, to secure state control over the marketing of wheat and other grains, and it is entirely in accord with the developments which have taken place in nearly all other wheat countries. . . ."[11]

Whatever preference the growers had for a wheat board over a voluntary co-operative—and the preference varied from time to time—the one insistent element in the western viewpoint after 1920 was the belief that the open market or speculative system was detrimental to the producers and therefore could not be tolerated. A national board with a monopoly of the wheat market meant the entire removal of the futures market and for that reason if for no other had much to commend it. A voluntary contract pool did not remove the futures market but it at least provided all growers with an alternative for the disposal of their crop.

The Conservative government under Prime Minister Bennett eventually yielded to the representations of western wheat growers who persistently urged the restoration of a national wheat board. The concession came, however, only with the prospect of a general election. Notice of motion of a wheat board bill was introduced into the Canadian House of Commons on March 4, 1935. The bill was introduced on June 19 and was referred to a special committee[12] for consideration. Reported back in substantially amended form the bill passed and was assented to on July 5. Mr. J. I. McFarland was appointed to the position of chief

[10]For the source of these excerpts and for a brief summation of the wheat board movement in Canada see Canadian Co-operative Wheat Producers Limited, *Directors' Report*, 1934–5, in Saskatchewan Co-operative Wheat Producers Limited, *Eleventh Annual Report*, 1935, pp. 54–8.

[11]*The Canadian Wheat Pools on the Air*, Second Series, March, 1936, p. 14.

[12]See Special Committee on Bill 98, Canadian Grain Board Act, *Minutes of Proceedings and Evidence*. Prime Minister Bennett was chairman of this committee.

commissioner. The Liberals, who were returned to office in the elections of October 14, 1935, removed McFarland and his colleagues at the beginning of December and appointed a new board with members of their own choice. Mr. J. R. Murray became chief commissioner on December 3, 1935.

A study of the evolution of federal wheat marketing policy contained in a single chapter can obviously include but little detail concerning the workings of the various agencies involved. In dealing with the new Wheat Board period which began in 1935 the significant features to be sought for are those relating to changes in policy, and description of Wheat Board operations is essential only in so far as it indicates or exemplifies such changes. Ample records of operational detail are to be found elsewhere.[13]

The re-establishment of the Canadian Wheat Board involved a major change in federal wheat marketing policy, but by no means the latest or most significant of such changes. A wheat board may have wide or narrow powers and responsibilities in relation to the market, and a legislative or executive alteration of the scope of such powers and responsibilities may represent a change of policy as fundamental as that expressed in the establishment of the board in the first place. These generalities are clearly exemplified in the case of the Canadian Wheat Board which at the present time has had two decades of continuous operation.

The bill as originally submitted to Parliament provided for an agency with the widest possible powers and responsibilities. It was, in fact, to be a *grain* board rather than a wheat board, with power to take over all elevators in the prairie provinces and to exercise exclusive control over the inter-provincial and export movement of the various grains grown in the prairie provinces. The new board was accordingly to possess a monopoly of the marketing of western Canadian wheat as had the board of 1919–20. It would also, however, have a monopoly of the marketing of coarse grains (oats, barley, and rye) and flaxseed, powers that the board of 1919–20 had not possessed. The bill, although envisaged as a temporary measure to give belated legislative structure and formality to McFarland's stabilization operations, was nevertheless drafted in terms of permanence. It proposed the absorption of all agencies for the purchase and sale of western grains by the projected board, a permanent

[13]See particularly T. W. Grindley *et al.*, *The Canadian Wheat Board, 1935–46* (Ottawa: King's Printer, 1947), reprinted from *Canada Year Book*, 1939 and 1947 editions.

national agency. As Dr. MacGibbon remarks, "This was a long step in the direction of state socialism for a Conservative government."[14]

Although willing to introduce a bill of such sweeping design, the Conservative government had not the courage or the conviction to force it through committee, in the face of the stubborn opposition which promptly developed, without making major compromises and alterations. The proposed grain board emerged as a wheat board. Instead of being the exclusive agency for the marketing of western wheat as foreshadowed in the bill, the board appeared in the final act as an alternative to the open market, to be selected or rejected at will by the individual grower. The compulsory features of the bill, which provided for the nationalization of the western elevator system as well, appeared in the act (ss. 9, 10, 11 and 16) subject to proclamation, but were not proclaimed. So long as these sections were not in effect, the board was merely required to establish annually a minimum price at which it would purchase wheat offered for sale by the grower, and to issue participation certificates which would entitle him to share in any additional proceeds. The individual grower might sell all, or none, or any intermediate portion of his wheat crop to the board. The board's fixed price provided a floor below which no grower needed to dispose of his crop. If sales by the board yielded returns above the price paid initially to the grower, the surplus was to be distributed on a pooled basis, grade by grade. If sales yielded less than the initial advance, the advance was nevertheless without recourse, and the shortage became a charge upon the federal treasury.

The wheat board legislation of 1935 fell short of meeting the demands of western wheat growers. Instead of providing for the cessation of futures trading in wheat in Canada, the act provided only for an optional marketing channel which freed producers from dependence on the open market system without in any way interfering with that system. Nevertheless, the decision of the government to provide an alternative to the open market for wheat was by no means unimportant for the evolution of agricultural policy. The legislation did not indicate the considerations which were to guide the board in its determination of the minimum fixed annual price, but once set, this price, subject only to the approval of the Governor in Council, carried the full financial guarantee of the Dominion government. The government had been drawn into stabilization operations of necessity and without formulating a specific principle. The board was designed partially as an agency to dispose of the accumulated stocks of stabilization wheat, but also to acquire and

14*The Canadian Grain Trade*, p. 35.

dispose of additional wheat. There was no intention that it should operate merely as a temporary agency for the disposal of existing stocks. Dr. Grindley has interpreted the wheat board legislation of 1935 as follows:

There is no doubt that the intent of the Canadian Wheat Board Act, 1935, was to protect the Canadian producer against untimely developments in the international wheat situation. In actual fact, the Canadian Wheat Board, through its power to fix a minimum price, through its power to receive Dominion financing, and through its power to transfer deficits to the Dominion Government, really acted as a buffer between chaotic conditions in the international wheat market and the farmers on the land in Western Canada.[15]

One of the very practical tests of the new legislation was the price policy pursued by the board. If the initial price established were so low as to be completely free from the danger of overpayment, the "buffer" effect of which Dr. Grindley speaks would be negligible.[16] If, on the other hand, the minimum price were deliberately set high in relation to market expectations, it would be possible to assure to the producer any minimum price which might be regarded as satisfactory. A third alternative, however, was for the board to forecast market probabilities at the beginning of the crop year and to establish the initial price as closely as possible in line with reasonable expectations. The latter appears to have been the approach adopted by the board in the early years of its operations, although other factors, including experience, led to repeated modification, not only in administrative but also in legislative provisions.

The board announced an initial payment of 87½ cents per bushel (basis No. 1 Northern, Fort William or Vancouver) on September 6, 1935, a price slightly above the open market price at the time. This price remained unchanged for the three crop years from 1935–6 to 1937–8, inclusive, but with considerable variation in results. During the crop year 1935–6 the board's price was above the market price for approximately eight months and the board took delivery of 151 million bushels of wheat or 70 per cent of total producers' marketings.[17] The

[15]*The Canadian Wheat Board*, pp. 7–8.

[16]Although the individual grower would still be protected from the possibility of sales on particular days at prices which might be below the annual average. This, of course, is because the board's sales returns are pooled.

[17]Producers made substantial deliveries to the board even when its price was below the market price. The board was selling stabilization wheat stocks along with those of the new crop year. At the end of the crop year the carryover comprised two million bushels of 1935 wheat and 82.7 million bushels of stabilization wheat. See Grindley, *The Canadian Wheat Board*, pp. 8–11.

final disposition of these stocks showed a net loss, or net overpayment to growers, of $11.9 million. The federal cabinet approved the price of 87½ cents per bushel for each of the two succeeding crop years (1936–7 and 1937–8) with the qualification that the board should not accept wheat unless the market price fell below 90 cents per bushel. Under these provisions the board did not handle any wheat from the crops of 1936 or 1937 and was operative only in completing the disposal of the stabilization stocks accumulated prior to 1935. The initial price as approved by the government for the crop year 1938–9 was 80 cents per bushel, basis No. 1 Northern, Fort William. Under extreme pressure from record world wheat production the market price fell below this figure and remained so throughout the entire crop year. The board received the entire marketed crop of 1938 (291.4 million bushels) and after final disposal of a substantial carryover, the board's net loss on the 1938 crop was placed at $61.5 million.[18]

The board had been in existence for approximately four years before the Liberal government faced any issue of fundamental policy in relation to it. The original Wheat Board Act of 1935 was a Conservative measure, modified in committee, it is true, to meet the opposition of various interests including that of the Liberals as a political group. The purge of personnel which followed the Liberal victory at the polls was made ostensibly to stimulate sales but without any major alteration of agency or policy. Stabilization stocks of wheat could not be liquidated overnight and the board represented an acceptable instrument for liquidating them over a period of time. A political fracas concerning the switch in the board's personnel led, in 1936, to the appointment of a special committee[19] of the House to investigate the operations of the Wheat Board, and this, in turn, led to the appointment of a royal commission to examine the problem of wheat marketing in all its aspects. Hon. Mr. Justice Turgeon, a veteran royal commissioner, was appointed to conduct this inquiry in 1936 but did not present his report until May, 1938. At the latter date the board had but recently closed out the stabilization account. All in all, the 1939 session of the House provided the first opportunity for the Liberal administration, elected in 1935, to express itself unequivocally concerning the existence of the Wheat Board.

The Liberals did not disband the Wheat Board in 1939 or change its purposes or policies to any significant degree. A superficial reading

[18]*Report of the Canadian Wheat Board, Crop Year 1941–1942*, p. 23.
[19]See Special Committee on the Marketing of Wheat and Other Grains under Guarantee of the Dominion Government, *Minutes of Proceedings and Evidence* (Ottawa, 1936).

of events would not, in fact, suggest that any major policy issues relating to the marketing of Canadian wheat had been dealt with in 1939 or at any time between the election of 1935 and the outbreak of the Second World War. A number of measures with specific reference to Canadian wheat were enacted in the session of 1939, but more than one of them remained inoperative.[20] The last pre-war year is not ordinarily regarded as one of formative significance in the development of Canadian marketing policy.

The fact that the Wheat Board emerged from the session of 1939 in but slightly modified form provides no indication of the stormy passage which it weathered in the legislature or of the compromise accepted by the administration under pressure from the western electorate. It was the settled intention of the government in the early weeks of the session to rid itself once and for all of the incubus of responsibility for the disposal of western wheat. The announcement of this intention, in terms which were as unequivocal as might be expected in political debate, called forth such a storm of protest from the West and provided so obvious a potential rallying point for disaffection in the general election, which could not be delayed beyond another year, that the government capitulated. A bill to provide for the continuance of the Wheat Board was introduced quietly and with an air designed to suggest that any other plan of action had never been even remotely contemplated. The record, however, effectively disposes of this fiction.

The Speech from the Throne, delivered on January 12, 1939, gave but the vaguest of hints regarding agricultural policy. The government pointed out that it had approved a fixed initial price for wheat as recommended by the Wheat Board for the crop of 1938–9 (80 cents per bushel, basis No. 1 Northern, Fort William), "in order to avert economic disaster to a large part of the population." It stated that the prairie farm rehabilitation plan would be continued and that legislation would be introduced to regulate grain exchanges as recommended by the royal commission on the grain trade (the Turgeon Commission of 1936–8) "and to assist further in the marketing of farm products." The debate on the Address in Reply added little information. In answer to comments and queries from opposition leaders, Mr. Gardiner, Minister of Agriculture, would say nothing about the government's intentions but stressed at some length the increase in wheat acreage which had taken place throughout the world since 1914. It was a matter of common surmise

[20]For a description and critical evaluation of this legislation see G. E. Britnell, "Dominion Legislation Affecting Western Agriculture, 1939," *Canadian Journal of Economics and Political Science*, VI, no. 2, May, 1940, pp. 275–82.

at the time that the initial price established for the crop of 1938 was likely to involve the federal treasury in a loss approximating thirty or forty million dollars at the least.

The Bren Gun discussion followed the debate on the Address in Reply, the two occupying more than a month. On February 14 the House, in committee of supply, turned to consider the estimates of the Department of Agriculture. Almost immediately the Minister of Agriculture was pressed to inform the House of the government's intentions with reference to the marketing of wheat. Mr. Gardiner pointed out that it would be quite improper for him to do so since all matters having to do with the marketing of wheat were the responsibility of the Department of Trade and Commerce rather than of the Department of Agriculture. Two days later, however, he took the opportunity offered by the supply discussions on the marketing service of his department to indicate in some detail the views and intentions of the government in regard to marketing wheat.

The general nature of the government's purposes regarding wheat could readily be detected very early in the Minister's speech although he reserved the revelation of specific intentions until near the end. He began by sketching developments in the Canadian grain trade after 1935 when the measure which emerged in modified form as the Wheat Board Act had been introduced by the Conservative government. He stressed that the act represented a compromise and referred repeatedly to what appeared to him as an obvious fact, that the measure had been originally intended exclusively for emergencies and had never been regarded as possessing any claim to permanence.[21] The initial price,

[21]The following excerpts from Mr. Gardiner's speech show clearly his attempt to establish the temporary, emergency character of the Wheat Board legislation of 1935 (in Canada, *House of Commons Debates*, Feb. 16, 1939, pp. 1033-7): "A compromise bill was finally agreed upon, which became the Wheat Board Act, 1935. The compromise act provides. . . . The discussions of the time would indicate that the final compromise legislation was intended to deal with an emergent condition resulting from national policies followed in exporting and importing countries. . . . Three years ago when the government came into office a wheat board had been set up under the Wheat Board Act of 1935. The government was of the opinion that this legislation only provided a means for dealing with an emergency created through the methods followed during previous years in attempting to market Canadian wheat under existing world conditions. . . . When the 1938 crop was coming on the market . . . it was considered that an emergency existed which, together with the difficulties experienced by the wheat producers in the years immediately preceding, necessitated the establishment of a minimum price which would in all probability pay a bonus to the wheat producing farmer. . . . It can be justified as the legislation has been justified from the beginning, namely, that it is dealing with an emergency situation. . . . We cannot agree that there is likely to be permanency to any system of marketing farm products which is based upon price fixing."

established annually by the government on the recommendation of the Wheat Board, had had either one of two possible results. If the board's price were below the market, as proved to be the case for the crop years 1936–7 and 1937–8, producers ignored the board and disposed of their wheat on the open market. If, on the other hand, the market price fell below that offered by the board and continued in that position, as it had done particularly since the autumn of 1938, producers took full advantage of the facilities of the board to secure the bonus assured them by an initial price, no part of which was recoverable by the government agency.

The overpayment or bonus emerging from the disposal of the crop of 1938 was, of course, still indeterminate but would amount, the Minister estimated, to 15 or 20 cents per bushel or a total of some $48 million. The grant of a bonus of this magnitude to western farmers might be justified, Mr. Gardiner argued, but only if the existing circumstances constituted an emergency. While the total might be justified, its regional and personal distribution could not. Since the bonus resulted from the establishment of a fixed initial price per bushel, the total payable to an individual farmer or within a particular area varied directly with the bounty of the crop and thus, the Minister argued, inversely in proportion to need. Of the three prairie provinces, the Minister pointed out, Saskatchewan was by far the most reliant on wheat production with an acreage exceeding that of the other two provinces combined. Yet, because of the regional incidence of complete and partial crop failures in 1938, Saskatchewan farmers would secure only $18 million out of the estimated total federal bonus of $48 million for the crop year. This type of award the Minister regarded as unjustifiable. He said: "I suggest to the committee [of supply] that a policy which gets results of that kind is not sound policy; that a policy which gives, unto those who have, still more and gives nothing at all to those who have nothing, is not sound policy to follow in connection with the marketing of wheat in this country."[22]

It is not difficult to imagine the conclusion of this line of reasoning or to foresee the type of policy declaration toward which the Minister of Agriculture was directing his analysis. He summarized his views, and presumably those of the government, regarding the Wheat Board, in the following words:

Three years of experience with the Wheat Board of 1935 has shown that it is legislation which can only be helpful to deal with a marketing emergency, and could not form the basis for a permanent system of marketing. The

[22]*Ibid.*, p. 1035.

present year has shown that it does not provide an equitable method of dealing with the emergency created through drought and grasshoppers. The minimum price per bushel results in those having most receiving most and those having no crop receiving nothing.[23]

As for the policy implications of these views, the Minister went on:

Realizing that the present legislation does not offer a solution for our marketing problems . . . we intend to introduce legislation which will carry out as far as possible the recommendations of the Turgeon commission, which were:

First, that the government should remain out of the grain trade and our wheat should be marketed by means of the futures market system.

Second, that the grain exchange should be placed under proper supervision.

Third, that encouragement be given to the creation of cooperative marketing associations or pools.

May I repeat that it is our intention to bring down legislation which we hope will make effective all of those recommendations.

The economic emergency which existed in the West could not, the Minister argued, be entirely disposed of by marketing legislation. It was necessary to reckon with the disastrous effects of "drought and grasshoppers." The Turgeon Commission had not been directed to consider these but Mr. Gardiner informed the House that the government intended to introduce legislation to deal with production failures, legislation which, the government hoped, would "make home building on the prairies more secure . . . and which would be drafted on the principle that assistance would be given in proportion to need, calculated on an acreage basis, and so adjusted as to encourage home building and maintenance rather than increased wheat production."[24]

The general and by no means unreasonable inference was that the Wheat Board was to be discontinued. Meanwhile, however, various groups in the West had been considering the wheat problem on the assumption that the board or some similar agency would continue to underwrite the price of the western staple product. Premier Bracken had convened an elaborate four-day conference in Winnipeg in December, 1938, to consider the question of markets for western farm products. The conference was attended by four hundred persons of whom many were official delegates of farmers' organizations, boards of trade, grain trade groups, and universities. The price of wheat was only one of the many topics discussed by the conference and the Wheat Board was accorded only the most incidental of references. Professor E. C. Hope, then of the University of Saskatchewan, estimated that the existing

[23]*Ibid.*, p. 1037.
[24]*Ibid.*

guaranteed price for wheat might cost the federal government $60 million for the crop of 1938 and yet was insufficient to allow farmers to pay interest and principal on their debt. In his opinion the guaranteed initial price provided a bonus which was unsatisfactory in principle because it varied directly with production and inversely with need. Nevertheless, Mr. L. W. Brockington, counsel for the North West Grain Dealers' Association, gave an unchallenged summary of the views of the conference in the following words:

. . . there is no argument whatsoever in the mind of any westerner against the maintenance of a just price for the western farmer, a price which guarantees him a decent standard of living. For that reason I think every organization in western Canada connected with the farmer's business this year did support a Wheat Board, and did support a price of not less than 80 cents, and, I believe, next year will support either a Wheat Board or whatever alternative can be set up to guarantee the farmer no less than a similar price to safeguard a decent living.[25]

On January 21, 1939, Premier Bracken named a continuing committee, the Western Committee on Markets and Agricultural Readjustment, in conformity with the directive of the Winnipeg conference and composed of twenty members representing particularly the agricultural organizations and governments of the prairie provinces. This group met in Regina on the same day and drew up a resolution which concluded as follows: "By way of general direction to its sub-committee, this committee offers for its guidance, its unanimous opinion that the Dominion government should be requested to extend to the western wheat growers in respect of the wheat to be grown in the crop year 1939–40 at least the protection given in the crop year 1938–39."[26] On February 14 the Western Committee met in Saskatoon; again it passed a resolution asking for continuation of the Wheat Board and for a fixed price of not less than 80 cents per bushel. This time it made specific arrangements for the presentation of its requests to the government and a delegation proceeded to Ottawa on March 1. Before the views of the delegation were laid before the government, however, the Minister of Agriculture had given advance notice of the government's wheat policy as outlined above.

Under pressure of the protest voiced by formal delegation and by representations to western members, the government gave up its intention to disband the Wheat Board. On March 27 Hon. Mr. Euler, Minister of

[25]*Proceedings of the Conference on Markets for Western Farm Products*, as arranged by the government of Manitoba and held in Winnipeg, Dec. 12 to 15 inclusive, 1938 (Winnipeg, Jan., 1939), p. 235.

[26]Canada, *House of Commons Debates*, May 10, 1939, p. 3848.

Trade and Commerce, gave first reading to Bill 63 to amend the Canadian Wheat Board Act, 1935. The bill contemplated leaving the board intact but proposed that its power to recommend an annual initial payment be replaced by the legislative provision for an initial payment of 60 cents per bushel, No. 1 Northern, Fort William.

The opposition to the newly proposed measure was greatly in excess of that which had greeted the prospect of the complete abandonment of the board. It was as if western constituents had perfect faith that no government in its right senses would disband the board and that even ministerial remarks which suggested such action must be the result of some oversight which would quickly be set to rights. The belated appearance of Bill 63 created no surprise in its assurance that the board would continue, but the clause providing for a statutory initial payment of 60 cents instead of 80 cents as then in effect caused the utmost consternation. The Western Committee sent a second delegation urging the continuance of the existing initial price. The legislatures of the prairie provinces forwarded resolutions. Western members were deluged with communications and petitions in support of the 80-cent initial payment. By the time Bill 63 came up for second reading on May 5, certain changes had been made. The initial payment stood at 70 instead of 60 cents but was still to be established on a statutory basis. A new feature of the bill was the provision that no individual farmer could deliver more than 5,000 bushels of wheat to the board within any one crop year.

The proposed initial payment was far from satisfactory to western groups in and out of the House, and the limitation on deliveries provided the basis for vigorous Opposition argument. The government, however, associated Bill 83, which embodied a plan for crop-failure assistance on an acreage basis,[27] with Bill 63, the Wheat Board amendment measure, for discussion purposes. Details of the crop-failure measure gave scope for criticism and debate but the proposal as a whole was acceptable without question. The government made it clear that it had gone as far as it had any intention of going in the new version of Bill 63 by way of compromise on the Wheat Board question. The reduction in the initial price, its establishment on a legislative basis, and the provision of a delivery quota, all those were to stand. Anyone who found these unsatisfactory was to keep in mind that the amendments to the Wheat Board Act were complemented by the new crop-failure or crop-insurance legislation. The government had a fully justified faith

[27]Bill 83 emerged as the Prairie Farm Assistance Act, *Statutes of Canada*, 3 Geo. VI, c. 36 (1939).

that the combined programme would carry the House. It also, apparently, regarded it as constituting a sound plank of western agricultural policy to build into the platform for the approaching election.

The continuance of the Wheat Board in 1939 cannot be taken as a matter of course. It involved a sharp division over declared governmental policy and, in the outcome, a major retreat on the part of the administration. Dr. MacGibbon has summarized the situation with clarity and conviction in the following words:

The really significant fact was that the Government had tried to get out of the business of selling wheat and had failed. Its inability to divest itself of the Canadian Wheat Board at this time registered the fact that the Canadian wheat growers were determined that the Government should maintain the Board, if not primarily as a regular vehicle for selling wheat, at least as a stand-by organization that would protect them against drastic downward swings in the market. The producers were firmly convinced that they were entitled to a "fair" price for their product, and if this was not forthcoming on the basis of supply and demand then they looked to the Government to make good the deficiency through the Wheat Board. With the capitulation of the Government to this point of view it was no longer possible to regard the Canadian Wheat Board as a temporary organization established to meet a single emergency, but as a permanent part of the institutional machinery of Canada for dealing with wheat marketing. . . .[28]

The early years of the Second World War provided no opportunity for the formulation of wheat marketing policy other than at the level of immediate contingency. The Wheat Board was retained as an optional marketing channel with no change in the initial payment of 70 cents per bushel for the crop years 1939–40 to 1941–2 inclusive. A delegation of four hundred wheat growers presented a petition bearing 185,000 names to the federal government in February, 1942, demanding that this price be increased to a minimum of $1.00. The government reluctantly conceded an increase to 90 cents. Meanwhile, exceptionally large wheat crops and the disappearance of European markets had led to the accumulation of unprecedented stocks of wheat in Canada. The Dominion government dealt with the emergency by introducing marketing quotas in 1940 and an acreage reduction programme in 1941. The reduction programme was continued for three years. Its purposes were, however, temporary, and the transfer of wheat acreage to the production of coarse grains which it effected was largely cancelled out by a transfer back again even before the end of the war. There is, in any case, no place in the present study for a detailed consideration of wartime policy except in so far as its effects have persisted into the post-war period.

[28]*The Canadian Grain Trade*, p. 47.

A significant change in marketing policy was made in 1943 when the Canadian Wheat Board ceased to be merely an alternative marketing channel and became the exclusive initial recipient of Canadian wheat as delivered from the farm. The Dominion government announced the new policy on September 27, 1943, to take effect immediately.[29] According to the new programme, trading in wheat futures was discontinued on the Winnipeg Grain Exchange; the Wheat Board was to acquire on behalf of the Dominion government all unsold wheat in commercial positions in Canada at the current market price; the board's initial payment was raised from 90 cents to $1.25 per bushel; and the crop accounts for the crop years 1940–1, 1941–2, and 1942–3 were to be closed out at the closing market price ($1.23½) as of the current date. The wheat acquired by the board under the terms of the new policy was primarily earmarked to meet requirements under Mutual Aid and to supply domestic purchasers. The new programme was originally projected to July 31, 1945.

The establishment of the Wheat Board as a monopoly in September, 1943, was a reaction to wartime emergency and did not represent matured convictions on the part of the federal government. The Canadian carryover of 424 million bushels as of July 31, 1942, had been only slightly reduced from that of the previous year, and to it was added one of the largest crops in Canadian history, that of 1942, totalling well over half a billion bushels. Market prospects were uncertain. In January, 1943, the Minister of Trade and Commerce announced a continuance of delivery quotas and of the 90-cent initial payment for the crop of 1943. On March 30 a government majority defeated a motion in the federal House that steps be taken to abolish futures trading in the Winnipeg Grain Exchange.

Meanwhile, however, the American supply of wheat had dropped sharply due to the unexpected disappearance of the cereal for food, livestock feed, and industrial conversion, with pronounced effects on prices in the grain markets of the western world. The price of spot wheat in the Winnipeg market held steady within a few cents of the initial Wheat Board price throughout the latter half of 1942 and the first quarter of 1943. At the end of March, 1943, the price of No. 1 Northern was slightly over $1.00 per bushel and it fluctuated closely around that figure until mid-July. Crops in the winter- and spring-wheat areas of

[29]The new wheat policy was set forth in detail under Order in Council P.C. 7942 dated October 12, 1943. See *Report of the Canadian Wheat Board, Crop Year 1943–1944*, p. 3. See also G. E. Britnell and V. C. Fowke, "Development of Wheat Marketing Policy in Canada," *Journal of Farm Economics*, XXXI, no. 4, Nov., 1949, pp. 627–42.

the United States were seriously impaired in the 1942–3 season and on April 30, 1943, the United States removed quotas on the import of Canadian wheat for feeding purposes. Canadian wheat acreage had been substantially reduced under the acreage reduction programme and the crop prospects, even on the reduced acreage, developed unfavourably throughout the summer of 1943. By the opening of the new crop year on August 1, open market prices were firmly established at levels which were 20 or 30 cents above the initial payment set by the board, with the result that deliveries were being made almost exclusively to the open market. The desire of the Dominion government to be assured of ample supplies of wheat to meet Mutual Aid commitments and to continue the practice of negotiating bulk sales of wheat to the British Cereal Imports Committee, coupled with serious concern for the domestic price structure, is sufficient, when related to the supply circumstances as outlined above, to explain the new wheat policy of September, 1943.

Dr. MacGibbon speaks of the situation in the early autumn of 1943 as one which "forced the hand" of the federal government. Elaborating his view he continues:

Although up to this time the Liberal administration had consistently displayed a theoretical bias in favour of the open market system of selling wheat, it found itself at length compelled to go the whole distance and to place sole responsibility upon the Canadian Wheat Board for disposing of Canada's wheat crop. A combination of causes had finally driven it into this position, rather than a tardy conversion to the belief that a national wheat board was the most desirable method of marketing the Canadian crop.[30]

It is not without significance that compulsory state marketing of Canadian wheat was instituted in 1943 in order to assure the availability of supplies and to arrest an advance rather than a decline in the price of wheat. These were the same central purposes which had led to governmental intervention in the grain trade during the First World War and to the establishment of the Wheat Board in 1919.

The monopoly assigned to the Canadian Wheat Board in 1943 has been maintained to the present time. Although private grain trade interests have supported a persistent campaign of publicity to inform the public of the merits of the open market system, the wheat producer has remained singularly unimpressed and has succeeded in preventing the government from regarding the restoration of the open market as a matter of practical politics. Early in the post-war period the necessity for state control of breadstuffs inherent in the Mutual Aid programme

[30]*The Canadian Grain Trade*, p. 64.

gave way to the equally urgent necessity for state control of breadstuffs in order to meet commitments under international bulk trading agreements. Two complementary steps in marketing policy were taken by the Dominion government in 1946 which, by their deliberateness, suggested that the government was in no hurry to seek the re-establishment of the pre-war wheat marketing mechanism. These steps were the negotiation of the United Kingdom Wheat Agreement and the creation of a five-year national pool for Canadian wheat.

The United Kingdom Wheat Agreement was announced July 24, 1946. According to its terms the Canadian government agreed to supply 600 million bushels of wheat to the United Kingdom over the four crop years 1946–7 to 1949–50, 160 million bushels for each of the first two years and 140 million bushels for each of the last two. The price was fixed at $1.55 per bushel, basis No. 1 Northern, Fort William, Vancouver, or Churchill for the first two years; for the third and fourth years the price was to be negotiated above minima of $1.25 and $1.00 per bushel for the two years respectively. Exports other than those to the United Kingdom would move at "world" prices with special controls applicable to sales to Canadian millers for domestic purposes. The domestic counterpart of the agreement was the creation of a five-year pool for all Canadian wheat, the pool made retroactive to include the crop of 1945 and designed to terminate on July 31, 1950, simultaneously with the United Kingdom Agreement. The producer's return from the pool was $1.833 per bushel, basis No. 1 Northern, Fort William, including a payment of 4.5 cents per bushel from the Dominion treasury.

International agreement, along with its domestic counterpart of monopoly over Canadian wheat marketings by the Canadian Wheat Board, continues. In March, 1949, the representatives of some forty countries signed the International Wheat Agreement in Washington to cover the four-year period from August 1, 1949, to July 31, 1953. Originally five exporting countries including Canada agreed to supply and thirty-six importing countries agreed to accept approximately 450 million bushels of wheat annually at prices falling within specific maximum and minimum figures. The maximum figure was $1.80 per bushel throughout the contract. The minimum figure was $1.50 per bushel for the first year and 10 cents per bushel less for each succeeding year. The minimum or floor prices, however, had no significance, for wheat sold at the maximum price for the duration of the agreement. For the final crop year of the contract, which ended July 31, 1953, forty-six signatory nations were under commitment to trade 581 million bushels. Of this commitment 97.6 per cent was met within the crop year. Canada's share in the agreement originally involved a commitment to supply 203 million bushels, an amount increased by the final year of the contract to 235 million bushels.

Throughout its first crop year the International Wheat Agreement ran concurrently with the Canada–United Kingdom Wheat Agreement and thus also with the final year of the five-year domestic pool. Since the conclusion of this pool on July 31, 1950, the Canadian Wheat Board has conducted an annual compulsory pool for Canadian wheat. The following can be stated concerning the return to the Canadian producer under the International Wheat Agreement. The crop year of 1949–50 yielded $1.833 per bushel, basis No. 1 Northern, Fort William–Port Arthur or Vancouver, the price, that is, of the five-year pool of which this crop year formed a part. The corresponding price for the 1950–1 crop year was $1.855; for 1951–2, $1.836; for 1952–3, $1.81872; for 1953–4, $1.56426; and for 1954–5, $1.65066.

A new International Wheat Agreement was secured in the spring of 1953. This agreement was to run for three years and the price range was between a maximum of $2.05 and a minimum of $1.55 per bushel. The new agreement was impaired by the refusal of the United Kingdom government to become a party to it at a maximum price above $2.00. Instead of embracing commitments covering approximately 600 million bushels of wheat annually the new agreement applies therefore to approximately 420 million bushels. Simultaneously with its approval of the new agreement in April, 1953, the Canadian government amended the Wheat Board Act of 1935 to extend the monopoly of the Canadian Wheat Board over Canadian wheat, oats, and barley to August 1, 1957. The board has controlled the marketing of western Canadian oats and barley since August 1, 1949.

PART FOUR

CONCLUSION

➤➤❈◀◀

CHAPTER FIFTEEN

The National Policy and the Wheat Economy

THE NATIONAL POLICY was defined in the early chapters of this study as a collective term covering those policies which after the middle of the nineteenth century were directed in complementary fashion toward the creation of a transcontinental Canadian nation. It has been recognized throughout the analysis that the making of a nation is more than politics and economics and that the diverse and intangible drives that go into the enterprise can be determined and evaluated only with the utmost difficulty. This study has nevertheless been almost exclusively economic and political because of the necessity of limiting it to manageable proportions. It has been argued that the origins and early development of the national policy can be traced back to the second quarter of the nineteenth century and that the end of its first hundred years coincided, in 1930, with the end of a distinct phase of Canadian economic and political history. The termination of this phase of the national policy left the federal government temporarily without a significant major purpose. The realization of this fact and the attempt to remedy it by reorientation of policies have concerned the Dominion government from time to time throughout the past quarter of a century.

Reorientation of policy has been the concern of the federal government only intermittently in recent decades because of the urgent requirements imposed by the depression of the thirties and the world war of the forties. The demands upon the federal government arising out of the depression were of such immediacy that little time was left for the formulation of long-range policy or, indeed, for recognition of the fact that traditional policies might require modification. The difficulties of the depression were not liquidated before the Munich pact of 1938 made clear the imminent certainty of European war. The war commenced within a year and the federal government directed its entire effort to the mobilization of Canadian defensive resources for participation in the conflict. Defence was, of course, not a new federal function but merely one of those assigned to the Dominion government by the British North America Act.

The national policy, directed toward the creation of a transcontinental Canadian nation, had as an immediate and critical requirement in its early stages the retention of the Pacific mainland and Vancouver Island. With the lower Columbia River and Puget Sound lost to the United States by the Oregon Boundary settlement, the possibility of locating satisfactory Pacific seaports in British American territory was seriously restricted. The Fraser River and its delta offered the opportunity for land-to-ocean contact but control of the river would be meaningless without possession of the massive barrier provided by Vancouver Island. Pacific mainland and flanking island alike, however, were separated from eastern British North American colonies by thousands of miles of territory, economically empty except for isolated fur-trade posts and the struggling Red River settlement. To hold and secure Pacific frontage for the British colonies it was essential that there be retention and effective occupation of the central plains. The economic imperative of the national policy, as contrasted with its political or territorial requirements, was the establishment of a new frontier of investment opportunity which would be attached commercially and financially to the eastern provinces. Although by no means clearly foreseen in the early decades of the national policy, effective occupation of the central plains as required for the preservation of Pacific frontage and a doorway to the Orient eventuated in the establishment of the wheat economy. This, in turn, provided the new economic frontier of investment opportunity without which the first century of the national policy would have been but an empty political dream.

It is particularly appropriate therefore to examine the interrelationships between the national policy and the wheat economy in historical perspective. As the threat of secession by the Pacific colonies abated before American inertia and the promise of a Canadian Pacific railway the agricultural prospect of the central plains assumed increasing importance in its own economic right. The Pacific railway would of necessity pass through the plains which lay between the mountains and Fort Garry and would link with eastern points beyond. The millions of empty acres in the vast expanse could not all be unfit for settlement or so poor as to be unattractive to land-starved Europeans. The "wheat economy" is a term that can be used with propriety only in retrospect for there was no time when the moulders of the national policy deliberately set about the task of building a regional way of life according to any such specific plan. Nevertheless, the establishment of a mid-continental agricultural economy which, it was clear, would have to rely heavily upon the production of wheat, came in the late nineteenth century to be the central economic aim within the broad purpose of the national policy. It is no mere coincidence that the relative completion of the processes creating the wheat economy in the late 1920's marked distinctly the end of a phase of the national policy and called for the discovery of new federal purposes or for the rediscovery of the old purposes in new form.

It is not difficult to recognize the success of the major objectives of the national policy in so far as they related to the West. The fact that the territory from the head of the Lakes to the Pacific is Canadian rather than American today and that the international boundary across the plains remains in the location to which it was assigned by treaty early in the nineteenth century is clear proof that the national policy did not fail in its first and perhaps greatest test. The vast reaches of the continent from the Great Lakes to the Pacific and the Arctic constituted the territorial objective both of the Americans, who in the mid-nineteenth century gathered aggressive inspiration from the concept of Manifest Destiny, and of the English interests who organized defensively within the national policy. The completion of the main line of the Canadian Pacific Railway in 1885 held the Pacific Coast and Vancouver Island and provided a firm and tangible declaration of intentions regarding the central plains. The construction of the Crow's Nest Pass branch in the 1890's made it possible to recover for Canada the Kootenay and other mountain valleys in which there was then developing the mineral empire of southern British Columbia. The establishment of the wheat economy constituted effective occupation of the plains and prevented American absorption of the comparatively limited agricultural region in the Canadian West and of the potentially significant stretches of northern Shield and forest as well.

Railway, land, and immigration policies were inextricably interrelated in the creation of the wheat economy. To the agricultural lands in the plains area fell the dual function of building railways and attracting settlers. The land policy of the federal government, administering western lands for the purposes of the Dominion, was consequently a compromise in which the free-homestead element, regarded uncritically as the ideal, was constantly subject to the prior claims of railway policy. The precedent established for the Canadian Pacific Railway Company that railway lands should all be fairly fit for settlement required the complementary concept of indemnity selection under which the railways could choose land any place in Manitoba or the territories, and the entire fertile zone became, in effect, a railway belt, a reservation from any part of which the railways might select earned acreage. The railway landgrant system came to an end in 1894 in the sense that no new agreements were made thereafter, but the process of earning by construction continued. Delay in construction was accompanied by still greater delay in selection so that it was not until 1908 that the last of the railway claims was liquidated and the location as well as the extent of these claims determined.

Railway lands were only one of the "for-sale" modifications of the free-homestead system in western Canada. Hudson's Bay Company and school lands, pre-emptions and purchased homesteads, lands sold directly by the Dominion government, irrigation lands, and miscellaneous other types all fell within the same category. Throughout the period of Dominion lands which terminated with the 1920's, the acreage disposed of in the prairie provinces by or for sale exceeded that disposed of under free and military homesteads and grants to half-breeds. An estimated total of 58.2 million acres had been disposed of by the various homestead and free-grant methods by 1928 while 61.3 million acres had been alienated by or for sale. Included in the latter category were 31.3 million acres of railway lands, 9.3 million acres of school lands, 7.0 million acres retained by the Hudson's Bay Company and 6.6 million acres of pre-emptions and purchased homesteads.

Although Dominion lands policy was a mixture of free-grant and for-sale elements, the ingredients and proportions of the mixture dictated for the most part by necessity rather than by choice, it is by no means clear that an unmixed system of any certain superiority could have been devised. The homestead unit of 160 acres, the quarter-section, was early and with increasing clarity demonstrated to be an uneconomically small acreage for agricultural production in substantial portions of the western plains. A free-grant unit of 320 acres might have been generally preferable although the evidence is by no means conclusively in favour of this alternative. With railway selection restricted to odd-numbered sections, and with Hudson's Bay Company and school lands distributed according to a specific pattern in units not larger than 640 acres, the result was that contiguous to every homestead quarter throughout a large part of the wheat economy there was another quarter-section available on a purchase basis. In the pre-emption area established to the west of Moose Jaw in 1908 the railways had declined to select land in any quantity so that the odd-numbered sections throughout this territory remained available for acquisition by the homesteaders by pre-emption or as purchased homesteads. The mixture of elements in Dominion lands policy contributed to the enlargement of productive units so essential to sound development in the wheat economy.

Few persons could be found today to argue that the national or even regional purposes in the West could have been better served by a land policy based exclusively on for-sale disposal. More readily encountered would be expressions of regret that the free-homestead segment of

Dominion policy could not have been expanded at the expense of the other methods of land alienation. As a corrective to any tendency to idealize the free-homestead system in retrospect it would be well to record some of its shortcomings. Assuming genuineness of motive, the filing of a homestead entry signified the candidate's intention of "proving up" and of securing title at the end of the three years' allotted time. While individual designs might be thwarted by exigencies entirely unrelated to the homestead process, including ordinary mortality, any substantial discrepancy between homestead entries and patents issued over a period of years would indicate failure in the realization of normal expectations. The discrepancy for the seventy years of Dominion lands administration is so pronounced as to indicate a wastefulness little less than shocking. Careful estimates place the acreage entered for on a homestead basis in western Canada in the period from 1870 to 1927 at 99 million acres and the corresponding patents, to 1930, at approximately 58.2 million. These data indicate a gap of over 40 per cent between expectation and fulfilment in the first critical phase of homesteading in Canada. In terms of human beings, four out of every ten Canadian homesteaders failed to "prove up," to secure title to their claim. The regional record was even worse. In Alberta, from 1905 to 1930, nearly 46 per cent of all homesteaders failed to prove up and, in Saskatchewan, in the period from 1911 to 1931, approximately 57 per cent—nearly six out of every ten—homesteaders abandoned their claims before securing title.

Receipt of title did not mark the end of hardship for the settler. Too often it only provided him with the collateral on which he could secure excessive credit or an amount of credit which, fully reasonable in a period of inflation, was wholly unrealistic in the succeeding periods of deflation and agricultural collapse. The difficulties associated with the misuse of credit and, more particularly, with fluctuations in the value of the dollar are not, of course, to be attributed to shortcomings in the homestead system or to any part of the land policies of the Dominion government. Available evidence, however, establishes the existence of one major failure in the over-all Dominion lands policy, the failure to base the settlement of the prairies on anything in the way of land or climatic surveys which would exclude from homestead entry those areas wholly unfitted for cultivation. The wholesale accumulation of resultant unwitting error on the part of western settlers is apparent in the concentration of abandoned farms in southwestern Saskatchewan and southeastern Alberta in the middle twenties and again throughout the thirties. The Prairie Farm Rehabilitation Act, passed by the Dominion government in 1935, was, in a sense, a further instrument of Dominion lands policy. The Administration established under the act has worked for twenty years with tremendous energy and enthusiasm, its efforts in substantial part devoted to correcting the mistakes of the homestead period.

Governments of the prairie provinces have had increasingly challenging responsibility during the past quarter of a century in attempting to formulate sound policy for the administration and development of a variety of natural resources, particularly those of the mine and the forest, with lesser emphasis on fisheries. Agricultural policy has continued to be of major concern to prairie governments since the transfer of natural resources as it was before, but only a small part of the problem has had to do with the alienation of agricultural lands. The simple reason for this is that comparatively little agricultural land remained unalienated in the prairie provinces by the time that the transfer to the provincial governments was effected. "One fact at least has become clear," said Professor Martin in 1938, "since the day of the railway land grant and free homestead a generation ago. The Dominion has disposed of the best agricultural lands of Western Canada."[1] And again, "By 1926 the agricultural lands of the Prairie Provinces had been substantially occupied. . . ."[2] By 1928 it was estimated that less than 12 per cent of the surveyed area remained to be disposed of in the prairie provinces.[3] In Alberta the proportion was 18 per cent and in Saskatchewan it was 4.6. The shortage of good agricultural land in the West had, in fact, become apparent many years before. As early as 1906 Sir Clifford Sifton stated that the government had "never stopped giving land grants [to railways] . . . until the land ran out."[4] The homestead and pre-emption era which began in 1908 was inaugurated after the practical disappearance of unalienated lands of a quality suitable for grant to railways.

The building of transcontinental railways and the settlement of western lands imposed responsibilities of such all-absorbing magnitude that only the military obligations which emerged with the outbreak of the First World War were sufficiently compelling to distract the attention of the Dominion government from the economic tasks of national development. The distraction was regarded as purely temporary. By the end of 1918 it was felt that the military job had been completed. The war was won. Canada had been defended and the world had been made

[1]Chester Martin, "Dominion Lands" Policy, Canadian Frontiers of Settlement, ed. W. A. Mackintosh and W. L. G. Joerg, vol. II, part II (Toronto, 1938), p. 495.
[2]Ibid., p. 519.
[3]Ibid., p. 495.
[4]As cited in ibid., pp. 329–30.

safe for democracy. The country divested itself decisively and completely of khaki and the Dominion government turned to pick up the threads of national development where they had been dropped in 1914.

By 1920, however, there was no longer any great project of national expansion based on western development waiting to be revived. Instead there were a number of comparatively uninspiring chores to be performed in order to bring the pre-war national project to a viable measure of completion. There were, for example, already too many transcontinental railways but there remained the frustrating task of salvaging two of them by co-ordination into a single system. The resultant administrative unit, the Canadian National Railways, was poorly equipped with feeder lines, and hundreds of miles of branches were added in the prairie region. The Dominion government added a terminal elevator at Prince Rupert by 1925; completed the Hudson Bay Railway and a terminal transfer elevator at Churchill by 1931; brought to completion the Welland Ship Canal which had been begun before the war, and built a five million bushel terminal elevator at Prescott to make the new canal effective for the movement of western wheat. Immigration was restored to substantial proportions in the late 1920's but comparatively few of the newcomers homesteaded in the prairie provinces. The restoration of the natural resources to the provincial governments was imminent and merely awaited the conclusion of a Dominion-provincial agreement.

The decade of the twenties was far from being one of economic stagnation in Canada. From 1920 to 1929 the population of the country increased by one-sixth, the real national income by one-half, and the volume of exports by three-quarters. New investment throughout the country exceeded $6 billion. The wheat economy expanded. From 1921 to 1931 the population of the prairie provinces increased by one-fifth, the area of occupied land increased by one-quarter, that of improved land by one-third, and the acreage of wheat by one-half. The tractor, truck, and combine-harvester appeared in significant numbers in the wheat growing region during the latter half of the decade. The expansion of the wheat economy, however, was overshadowed by the expansion of non-agricultural activities and the real capital formation associated therewith. The installation of massive new hydro-electric equipment was accompanied by a corresponding expansion of the equipment for the production of pulp and paper and for base metal mining. From 1920 to 1929 the developed waterpower in Canada increased from 2.5 million to 5.7 million horsepower, the gross value of pulp and paper production increased from $151 million to $244 million, and that of non-ferrous metals from $76 million to $154 million. The Canadian gold mining industry expanded apace during the thirties when all other types of productive activity were seriously depressed. The industries which expanded at maximum rates during the first inter-war decade and which displayed a corresponding ability to absorb and to attract new capital were concentrated in the central provinces and in British Columbia. They involved the utilization of provincial natural resources and the active promotion of these developments was a challenging responsibility for the respective provincial governments. The stature of the provincial governments rose steadily as compared with that of the Dominion throughout the twenties.

The transfer of natural resources to the governments of the prairie provinces was important as a symbol rather than as a cause of the practical if temporary disappearance of Dominion economic purpose. The developmental activities of the Dominion government had related almost exclusively to the settlement of the western territories and, by 1930, so little land of good agricultural quality remained unalienated in the prairie provinces that its transfer to provincial jurisdiction affected Dominion possibilities to a negligible extent. The prairie provincial governments hastened to organize departments and agencies for the administration of their newly acquired public domain but the accomplishment of further agricultural expansion of significant magnitude defied the best efforts of the provincial departments.

The passing of significant possibilities for further agricultural settlement interfered so sharply with federal purposes that a major reorientation in policy appeared to be imperative. In a previous publication[5] this interruption of federal purposes was interpreted as marking the end and fulfilment of the first national policy, and as setting the stage for a new or second such policy. However, evidence accumulates with increasing force to indicate that terms such as fulfilment and termination are too decisive to describe accurately the modifications in Dominion policy which centred upon 1930, that it is too soon to seek the transformation of a completed "first" national policy into an emergent second. The conclusion of a phase has been clearly established, the phase in which Dominion peacetime purpose was almost exclusively intent upon the furtherance of western agricultural expansion. The developmental purpose has, however, been restored in recent years, although in non-agricultural areas. This restoration will be commented on below.

Although the national policy was interrupted rather than terminated by the practical completion of the processes which established the wheat economy, the interruption was sufficient in intensity and duration to

[5]See V. C. Fowke, "The National Policy—Old and New," *Canadian Journal of Economics and Political Science*, XVIII, no. 3, Aug., 1952, pp. 271–86.

suggest the likelihood of substantial modification in federal outlook and design. The portrayal of the development of the wheat economy within the framework of the national policy has called for special attention to the agricultural segment of the economy and to the relations between agriculture and commerce in particular. There is justification, therefore, for concluding with a brief examination of the agricultural policy of the federal government in order to determine whether there are any changes in that policy and, if so, whether any of the changes promise a degree of permanence beyond that of immediate expediency. Attention may, for this purpose, remain centred upon the wheat economy, for federal agricultural policy has traditionally concerned itself with western agriculture to such a degree that a survey of policy in this area will serve to suggest trends if not to enumerate details.

The essential features of Canadian agricultural policy grow by direct implication out of underlying economic philosophy. It is a question of the place of agriculture within the price system and of the views of government concerning the maintenance or alteration of that place. The economic liberalism underlying Canadian governmental policy has maintained an effective barrier against proposals designed to alter the position of the agricultural producer within the price system. This is not to suggest that economic liberalism has been an unchanging guide to Canadian policy, or that there has ordinarily been close harmony between liberal fact and liberal fiction. Even the most liberal and "free-enterprising" of Canadian governments in the nineteenth century regarded the marshalling of resources for developmental purposes to be one of the essential functions of government. The provision of transportation facilities in this vast and inaccessible new country could not, it was realized, be entrusted to the prompting of prospects for profit. Roadways, waterways, and railways were inevitably costly in terms of capital and labour, they were indispensable to economic development, and their provision had therefore to be assured by all the resources of government.

The economic philosophy which underlay the national policy, at least until the end of its first major period of achievement in 1930, rationalized governmental enterprise and assistance of a developmental nature, governmental activity of a regulatory nature, and state-financed research in the field of production, but little more. Production and marketing, it was taken for granted, ought normally to be guided by the search for profits within the system of free enterprise. The term, "price system," was assumed to be synonymous with "competitive system." If monopolistic elements persisted in appearing they might nevertheless be disregarded as peripheral and accidental phenomena. Publicity rather than regulation would serve to restrain monopolists and to minimize their significance, for business men were men of good will with a high regard for their own good names. There was no place in this basic philosophy for trust-busting or even for any serious measure of muck-raking. The possibility of instituting public enterprise as a curb and counterpoise to private monopoly was unacceptable. The philosophy was consistent in that it justified equality of freedom to competitive and monopolistic entrepreneurs alike.

Monopoly, seldom existent in the absolute, is best regarded as a relative and variable measure of interference with the freedom of competition. It is a condition inherent in the search for profits. The profits in a competitive system accrue, paradoxically, to the monopolistic elements in the system or, in other words, vary inversely with the degree of competition and directly with the degree of monopoly. The possibilities of interfering with competition, in turn, vary with the number of producers in a field. Interference is easy where there are few producers and difficult where there are many. The inevitability of competition among agricultural producers and consumers because of the atomistic nature of agricultural production and the resultant multiplicity of agricultural producers has been indicated in earlier chapters. The facility with which the rigours of competition may be avoided by the other parties in markets where farmers buy and sell has also been indicated. The inequality of the bargaining strength of agricultural producers as compared with that of the groups to whom farmers sell and from whom they buy is the inevitable complement of freedom of enterprise which accords equal tolerance to freedom of combination and freedom of competition.

One of the most significant features of the national policy has been a persistent disregard of the competitive inferiority of agriculture within the price system. The major era of the national policy which ended in 1930 witnessed no serious attempt on the part of government to ameliorate or even to assess that inferiority. The question raised at this point is whether the agricultural policy of the federal government over the past quarter of a century has shown any modification on this score or whether it indicates merely a variation in detail without alteration in design. The answer can be given only tentatively here, for an exhaustive analysis of recent and current governmental policy is inappropriate for the concluding stages of the present study. The main elements of this policy can, however, be noted with a view to suggesting a reasonable answer to the question.

Two comparatively substantial segments of federal agricultural policy —research and regulation—represent long-range continuity rather than novelty. The areas of investigation and, to a lesser extent, those of regulation have been multiplied but the purposes of both are of long standing. Agricultural research and the governmental financing and direction of such research are based on the theory that the increase in agricultural productivity and the reduction of agricultural costs constitute a valid public contribution to the welfare of the agricultural community. Regulation such as that provided for in the Canada Grain Act provides marketing assistance by means of the definition and assurance of uniformities in weights and measures, qualities, and a considerable range of economic conduct.

The view that agriculture can best be assisted by measures which increase agricultural productivity is so widely and uncritically accepted that it requires comment. The highly competitive conditions under which agricultural products are brought to market have particular significance at this point. Under competitive circumstances the inevitable tendency in any period other than a very short one is toward equality between selling price and costs of production. Improvements in cultural practices, in methods of controlling pests or obtaining a greater variety of seeds or animals, which result from public research and are therefore accessible to all producers on equal terms, will consequently tend to reduce costs of production and selling prices in equal measure. Cost reductions in agriculture are quickly passed on to consumers by way of price reductions. The farmer shares with other consumers in the advantages of cheap and abundant foodstuffs, but he gains as a consumer instead of as a producer, and in proportion to his consumption of foodstuffs rather than in proportion to his productive efforts. There is little need to justify the use of the taxpayer's dollar for the support of research directed toward the reduction of food costs. To charge the agricultural research budget of government to the account of the agricultural producer is, however, scarcely legitimate.

It may be argued that the reasoning of the above paragraph applies primarily if not exclusively to the case of agricultural producers whose output is consumed entirely within the domestic market, and that the allocation of the gains from research and standardization is not so easily determined where the producers in question market their produce abroad as well as at home. The latter case typifies the Canadian wheat grower and is thus particularly relevant here. In such a case the result of effective research is to enable the producer to produce more cheaply and to sell more cheaply in both home and foreign markets. Domestic consumers will benefit through cheaper foodstuffs, and the agricultural producers of the country may be enabled to hold a larger share of the international market than they could otherwise have done. The agricultural output of the country will be larger than would otherwise be the case. Agricultural profits will tend to disappear under competitive pressure among domestic agricultural producers, but rents and the price of agricultural land may rise or be maintained, to the advantage of the nation.

Emphasis on the importance of the competitive nature of agricultural production makes it less difficult to assess the degree of significant novelty in recent agricultural policy. Among the more important of the specific measures representing this policy as far as it concerns western Canada are the following: the provision of credit under the Canadian Farm Loan Act of 1927 and the Farm Improvement Loans Act of 1944; the Prairie Farm Rehabilitation Act of 1935; the Prairie Farm Assistance Act of 1939; the Agricultural Prices Support Act of 1944 and the removal of the tariff on agricultural implements the same year; and the Wheat Board Act of 1935 as repeatedly amended. It is more important to consider whether or not these various measures have tended toward a restraint of the competition inherent in agriculture than it is to elaborate their specific functions and methods of operation.

Generally speaking, the measures listed above show slight regard for what might be called the competitive excesses of agriculture. They are for the most part patterned after the two-blades-of-grass type of assistance so commonly extended to agriculture, the assumption being, apparently, that if agricultural output could but be doubled the farmer would be twice as well off as before. State subsidization of facilities for agricultural credit is thus designed to lower the cost and to increase the availability of capital for farming operations. Assistance extended under the Farm Loan Act and Farm Improvement Loans Act is offered impartially to agricultural producers whether they are interested exclusively in the domestic market or in the export market as well. In the latter instance the assistance may create a national agricultural advantage which would accrue to landowners in the form of rents and enhanced values for agricultural land. The Prairie Farm Rehabilitation Administration, executing the most ambitious of all the phases of current Canadian agricultural policy, has worked imaginatively and with great enthusiasm toward the mastery of various production problems in the wheat economy. The wheat economy has thus been maintained as a more substantial segment of the Canadian economy than would otherwise have been the case. Neither the agricultural credit acts nor the rehabilitation legislation have the purpose or the result of modifying the competitive status of Canadian agricultural producers in domestic markets. The cost reductions attributable to governmental assistance under these statutes tend to be passed rapidly on to Canadian and foreign consumers instead of being retained as profits by agricultural producers.

The Prairie Farm Assistance Act provides a primitive and poorly rationalized form of state-subsidized income insurance to western Canadian wheat growers. The act imposes a levy of 1 per cent of the sale value of all wheat, oats, barley, and rye marketed in the prairie provinces, the proceeds to be segregated into a special governmental account for the purposes of the act. Under certain kinds of crop failure the individual grower is entitled to cash assistance to an amount not exceeding $2.50 per acre on one-half of his cultivated acreage and in no case exceeding $500 per year. Deficiencies between levies and payments are made up out of the Consolidated Revenue Fund. Table XI summarizes the financial history of the Prairie Farm Assistance programme over the fifteen crop years 1939–40 to 1953–4 inclusive.

TABLE XI

LEVIES, AWARDS, AND PAYMENTS UNDER THE PRAIRIE FARM ASSISTANCE ACT, 1939–40 TO 1953–4

	Levy of 1 per cent	Number of awards	Payments
Manitoba	$12,877,116	33,337	$ 3,985,815
Saskatchewan	46,542,465	457,311	104,997,981
Alberta	23,778,507	195,113	36,943,669
British Columbia	—	2,124	290,317
Unallocated	14,586		
TOTALS	$83,212,674	687,885	$146,217,782

SOURCE: Canadian Co-operative Wheat Producers Limited, *Directors' Report*, 1952–3, in Saskatchewan Co-operative Producers Limited, *Twenty-Ninth Annual Report*, 1953, p. 86, and *Thirtieth Annual Report*, 1954, p. 93.

A number of factors suggest a large measure of irrationality in the Prairie Farm Assistance programme as at present constituted. The original rate of levy and the rates of payment remain substantially unchanged after fifteen years despite the clear demonstration that they bear nothing but a purely fortuitous relationship to one another. The maximum payment of $500 in any year to an individual farmer was an appreciable sum in 1939 but has shrunk by approximately one-half in real terms through the governmentally tolerated inflation of the past twelve or fifteen years. If the acreage payment or the maximum payment per farmer had any other than arbitrary significance in 1939, neither can be regarded as anything but wholly unrealistic today.

The regional disparities in the relative incidence of levies and awards under the Prairie Farm Assistance programme are not supported by clear or persuasive justification. While the programme has been in effect, Manitoba farmers have *paid* $3.25 in levies for each dollar they have received in payments, whereas Saskatchewan farmers have *received* $2.30 and Alberta farmers $1.55 for each dollar they have paid in levies. The detailed regional records of the programme would no doubt indicate more striking and even less justifiable anomalies. One wonders about the extent and nature of the marginal fringe which attaches to the wheat economy by virtue of the individual prospect of a maximum cash subsidy of $500 per year.

The Agricultural Prices Support Act and the Wheat Board Act are the only ones in the group mentioned above which clearly imply a recognition of the hazards confronting agricultural producers because of unbridled competition. The former act was designed to provide governmental support for the prices of farm produce other than wheat. The support might be extended by outright purchase and sale by a government agency or by way of deficiency or equalization payments. The measure was originally intended to be merely temporary and was designed to facilitate the transition from war to peace. It was repeatedly re-enacted, however, and in 1950 was made permanent. The Dominion government, through the Agricultural Prices Support Board, has acted in specific years to support the prices of products such as potatoes, apples, butter, honey, dried skim milk, dried fruits, and dried beans. In the period from 1946, when the act was put into operation, to the end of 1951 the support measures conducted under its authority had cost the government approximately $10 million.

The adoption and maintenance of agricultural price support and wheat board legislation imply the existence in the legislative mind of at least a doubt regarding the universal efficacy of complete freedom of enterprise in agricultural markets. True, there is little assurance of more than doubt. The price support legislation was introduced as a temporary expedient and as a means of maintaining full employment. The maintenance of farm income, toward which it was expected to contribute, was regarded as desirable not so much as an aid to agriculture but in order that by this and other instruments such as family allowances the anticipated post-war depression might be averted. The Wheat Board was retained in 1939 on the basis of political realities rather than because of economic conviction, much less conversion.

The facts of today must, however, be noted also. Price support legislation designed for the period of transition was not allowed to lapse but instead was made permanent five years after the end of the war. The Wheat Board, established as a monopoly in the marketing of Canadian wheat in 1943, has had its scope repeatedly extended as the indispensable counterpart of a series of international bulk-trading agreements. The first of these, the Canada–United Kingdom Wheat Agreement of 1946, was bilateral, whereas the international wheat agreements of 1949 and 1953 were multilateral. The monopoly of the Wheat Board in the marketing of Canadian wheat has been extended to August 1, 1957, a year beyond the expiry date of the current International Wheat Agreement. The powers of the board have also been extended horizontally to include control over the marketing of oats and barley produced in western Canada. These powers became effective August 1, 1949, and are at present extended along with the board's monopoly over wheat to August 1, 1957.

The facts outlined in the preceding paragraph suggest that a new element may have found a place among the goals of Dominion agricultural policy—a willingness on the part of the government to interfere in agricultural markets in order to mitigate the worst of the effects of unbridled competition among agricultural producers. This is the clear implication of agricultural prices support and Wheat Board legislation and of the continuance of this legislation well beyond the post-war rehabilitation period. As long as this or similar legislation is on the statute books and in active use, the competitive position of agriculture within the Canadian price system is not exclusively an open market phenomenon. The question which comes to mind is, however, how long is legislation of this sort likely to remain on the Canadian statute books and in active use? Can it be assumed that the Dominion government is so firmly seized of the inequality of the agriculturist's bargaining power that the removal of this inequality may be regarded as one of the long-run goals of future agricultural policy? Here we are in the realm of surmise where the tentative results of deeper probing may not warrant the effort. A few brief comments nevertheless appear to be in order.

One may well doubt that Dominion agricultural policy is inspired by a conviction of the competitive inadequacy of agriculture. Mention has already been made of the fact that the Agricultural Prices Support Act was originally regarded as one of the guarantors of post-war employment. The Wheat Board, adopted to formalize stabilization operations and to interpose a buffer between the wheat producer and the possible chaos of world wheat markets, was retained in 1939 under direct and indirect political pressure rather than out of government conviction. The board was given the status of a monopoly in 1943 primarily as a measure of protection for the European consumers against anticipated scarcity and an advance in prices. This status has the continuing and general support of western wheat growers whose spokesmen stress its stabilizing influence when combined with international negotiation and bulk-trading agreements. These also have impressive support from the growers. The international agreements may have more than one desirable attribute for the national government. They may be regarded as a means toward the stabilization of the wheat economy and toward the maintenance of a high level of economic activity in the Dominion as a whole. Furthermore, the monopoly over, and the exclusive bargaining rights in, the annual supply of from seven to ten million tons of high-grade breadstuffs cannot be regarded as a matter of complete indifference to a government with comparatively new-found stature within the family of nations.

The persistence of doubt concerning the degree of permanence in Dominion agricultural policy arises mainly from its lack of theoretical or conceptual content. One may well have no liking for the concept of parity price and nevertheless feel impelled to recognize its effectiveness as a unifying force in an American farm policy which has great tenacity and a substantial measure of consistency. An examination of the development of the Canadian wheat economy within the framework of the national policy suggests that clichés concerning the welfare of agriculture have provided the rationalization rather than the reason for agricultural policy while the reasons themselves have been of such diversity and remoteness that consistency of policy has been impossible. It is suggested here that the prerequisite for consistency would be a clear recognition of the competitive disabilities of agriculture within the price system and a clear decision as to whether these disabilities are to be tolerated or removed.

Developments of very recent years suggest that Dominion policy may not require the fundamental reorientation that appeared so necessary throughout a great part of the second quarter of the present century. The prospects for federal intervention in national economic development are vastly greater today than they were a decade or two ago. The new circumstance is that the field of development is non-agricultural. Natural resources in the Yukon and the Northwest Territories remain under federal jurisdiction and a number of technological and geographic discoveries including those relating to uranium point to a developmental task of great if uncertain magnitude. The natural resources within the provinces are all within provincial control but their development has raised some of the most crucial of the national problems of the post-war years. Thus, for example, there is the question of the export or non-export of oil and gas from the prairie provinces, and questions concerning the location and the proper time of construction of transportation facilities for these products in order to foster the industrial development and urbanization of central Canada. The preservation of the east–west axis of trade and transportation is as urgent a requirement today as at any time in the past. The last ironic touch, and the one which finally suggests that the national policy may have turned full circle for a second start, is the recurrence of the problem of the improvement of the St. Lawrence waterway system. The development of iron ore deposits and waterpower resources on the Quebec-Labrador boundary may be matters of unquestioned provincial concern. The enlargement of the St. Lawrence seaway in order to funnel this ore into the industrial complex of the St. Lawrence Valley is a federal matter.

Bibliography

BOOKS

BLACK, J. D., *Agricultural Reform in the United States* (New York, 1929).
BUCKLEY, K. A. H., *Capital Formation in Canada, 1896–1930*, Canadian Studies in Economics, no. 2 (Toronto, 1955).
BURT, A. L., *The Old Province of Quebec* (Toronto, 1933).
CLOKIE, H. McD., *Canadian Government and Politics* (new and revised ed., Toronto, 1950).
COWAN, HELEN I., *British Emigration to British North America, 1783–1837* (Toronto, 1928).
CREIGHTON, D. G., *British North America at Confederation* (Ottawa, 1939).
—— *The Commercial Empire of the St. Lawrence, 1760–1850* (Toronto, 1937).
DAFOE, JOHN W., *Clifford Sifton in Relation to His Times* (Toronto, 1931).
DAVISSON, WALTER P., *Pooling Wheat in Canada* (Ottawa, 1927).
ELSWORTH, R. H., *Statistics of Farmers' Cooperative Business Organization* (Washington: Farm Credit Administration, 1936).
FILLEY, H. C., *Cooperation in Agriculture* (New York, 1929).
FOWKE, V. C., *Canadian Agricultural Policy: The Historical Pattern* (Toronto, 1946).
—— *The Purposes, Origins, and Development of the Canadian Pacific Railway Company as a Corporate Entity: Evidence in the Matter of an Application by the Canadian Pacific Railway Company for the Establishment of a Rate Base and Rate of Return before the Board of Transport Commissioners for Canada, January 5, 1953* (Regina, 1953).
GLAZEBROOK, G. P. DE T., *A History of Transportation in Canada* (Toronto, 1938).
GRINDLEY, T. W., *et al.*, *The Canadian Wheat Board, 1935–46* (Ottawa: King's Printer, 1947), reprinted from *Canada Year Book*, 1939 and 1947 eds.
HANSEN, A. H., *Fiscal Policy and Business Cycles* (New York, 1941).
HEDGES, JAMES B., *The Federal Railway Land Subsidy Policy of Canada* (Cambridge, Mass., 1934).
HENRY, LORNE J., and GILBERT PATERSON, *Pioneer Days in Ontario* (Toronto, 1938).
HOFFMAN, G. WRIGHT, *Futures Trading upon Organized Commodity Markets in the United States* (Philadelphia, 1932).
HOWAY, F. W., W. N. SAGE, and H. F. ANGUS, *British Columbia and the United States* (Toronto, 1942).
INNIS, H. A., *A History of the Canadian Pacific Railway* (Toronto, 1923).
—— ed., *The Diary of Alexander James McPhail* (Toronto, 1940).
—— and A. R. M. LOWER, eds., *Select Documents in Canadian Economic History, 1783–1835* (Toronto, 1933).
JAMESON, ANNA, *Winter Studies and Summer Rambles* (Toronto, reprint 1943).
LIPSET, S. M., *Agrarian Socialism: The Co-operative Commonwealth Federation in Saskatchewan* (Berkeley and Los Angeles, 1950).
LOWER, A. R. M., *Colony to Nation* (Toronto, 1946).

MacGibbon, D. A., *The Canadian Grain Trade* (Toronto, 1932).
—— *The Canadian Grain Trade, 1931–1951* (Toronto, 1952).
MacGregor, John, *British America* (2nd ed., Edinburgh, 1833).
Mackintosh, W. A., *Agricultural Cooperation in Western Canada* (Kingston, 1924).
—— *The Economic Background of Dominion-Provincial Relations* (Ottawa: King's Printer, 1939).
—— *Economic Problems of the Prairie Provinces* (Toronto, 1935).
—— and W. L. G. Joerg, eds., Canadian Frontiers of Settlement Series (Toronto: The Macmillan Company of Canada).
Macpherson, C. B., *Democracy in Alberta* (Toronto, 1953).
Magill, R., *Grain Inspection in Canada* (Ottawa, 1914).
Martin, Chester, *"Dominion Lands" Policy*, Canadian Frontiers of Settlement, ed. W. A. Mackintosh and W. L. G. Joerg, vol. II, part II (Toronto, 1938).
Masters, Donald C., *The Reciprocity Treaty of 1854* (London, 1936).
Mathieson, George S., *Wheat and the Futures Market: A Study of the Winnipeg Grain Exchange* (Winnipeg, 1942).
Mills, Joseph C., "A Study of the Canadian Council of Agriculture," unpublished M.A. thesis, University of Manitoba, 1949.
Morton, W. L., *The Progressive Party in Canada* (Toronto, 1950).
Patton, H. S., *Grain Growers' Coöperation in Western Canada* (Cambridge, Mass., 1928).
Pope, Joseph, *Correspondence of Sir John Macdonald* (New York, 1921).
Porritt, Edward, *Sixty Years of Protection in Canada, 1846–1912* (Winnipeg, 1913).
Pritchett, J. P., *The Red River Valley, 1811–1849* (New Haven, Conn., 1942).
Regier, C. C., *The Era of the Muckrakers* (Chapel Hill, N.C., 1932).
Reynolds, Lloyd G., *The Control of Competition in Canada* (Cambridge, Mass., 1940).
Seager, Henry R., and Charles A. Gulick, Jr., *Trust and Corporation Problems* (New York, 1929).
Skelton, O. D., *The Railway Builders* (Toronto, 1916).
Surface, Frank M., *The Grain Trade during the World War* (New York, 1928).
—— *The Stabilization of the Price of Wheat during the War and Its Effect upon the Returns to the Producer* (Washington: United States Grain Corporation, 1925).
Trotter, R. G., *Canadian Federation* (Toronto, 1924).
United Grain Growers Limited, *The Grain Growers Record, 1906 to 1943* (Winnipeg, 1944).
Webb, Walter Prescott, *The Great Plains* (Boston, 1931).
Weinberg, Albert K., *Manifest Destiny: A study of Nationalist Expansionism in American History* (Baltimore, 1935).
Yates, S. W., *The Saskatchewan Wheat Pool*, ed. Arthur S. Morton (Saskatoon: United Farmers of Canada, Saskatchewan Section, Limited, 1947).

Articles

Britnell, G. E., "Dominion Legislation Affecting Western Agriculture, 1939," *Canadian Journal of Economics and Political Science*, vol. VI, no. 2, May, 1940, pp. 275–82.
—— and V. C. Fowke, "Development of Wheat Marketing Policy in Canada," *Journal of Farm Economics*, vol. XXXI, no. 4, Nov., 1949, pp. 627–42.
Fowke, V. C., "Developments in Canadian Co-operation," *Journal of Farm Economics*, Proceedings Number, Nov., 1951, pp. 909–17.
—— "The National Policy—Old and New," *Canadian Journal of Economics and Political Science*, vol. XVIII, no. 3, Aug., 1952, pp. 271–86.

—— "Royal Commissions and Canadian Agricultural Policy," *ibid.*, vol. XIV, no. 2, May, 1948, pp. 163–75.

HIND, E. CORA, "A Story of Wheat," reprinted from *Canadian Geographic Journal*, Feb., 1931.

INNIS, H. A., "Unused Capacity as a Factor in Canadian Economic History," *Canadian Journal of Economics and Political Science*, vol. II, no. 1, Feb., 1936, pp. 1–15.

JAMES, C. C., "History of Farming" in A. Shortt and A. G. Doughty, eds., *Canada and Its Provinces* (Toronto, 1914–17), vol. XVIII.

MCLEAN, S. J., "National Highways Overland" in A. Shortt and A. G. Doughty, eds., *Canada and Its Provinces* (Toronto, 1914–17), vol. X.

PATTON, H. S., "The Canadian Wheat Pools in Prosperity and Depression" in Norman E. Himes, ed., *Economics, Sociology and the Modern World: Essays in Honor of T. N. Carver* (Cambridge, Mass., 1935).

PENTLAND, H. C., "The Role of Capital in Canadian Economic Development before 1875," *Canadian Journal of Economics and Political Science*, vol. XVI, no. 4, Nov., 1950, pp. 457–74.

SHARP, MITCHELL W., "Allied Wheat Buying in Relationship to Canadian Marketing Policy, 1914–18," *Canadian Journal of Economics and Political Science*, vol. VI, no. 3, Aug., 1940, pp. 372–89.

SKELTON, O. D., "General Economic History of the Dominion, 1867–1912" in A. Shortt and A. G. Doughty, eds., *Canada and Its Provinces* (Toronto, 1914–17), vol. IX.

—— "The Grain Markets of Britain" in Midland Bank Limited, *Monthly Review*, July–Aug., 1930, pp. 4–7; Aug.–Sept., 1930, pp. 5–8.

GOVERNMENT DOCUMENTS

Province of Canada

Parliamentary Debates on the Subject of the Confederation of the British North American Provinces, 3rd session, 8th provincial Parliament of Canada (Quebec, 1865).

Sessional Papers, 1862.

Canada

Board of Grain Commissioners, *Hand Book on the Sale and Handling of Grain through a Country Elevator* (n.d., n.p.).

Bureau of Statistics, *Wheat Review* (Ottawa: monthly).

"Historical Analysis of the Crow's Nest Pass Agreement and Rates," Part I of *Crow's Nest Pass Rates on Grain and Grain Products: Joint Submission of the Governments of Alberta, Saskatchewan, and Manitoba to the Royal Commission on Transportation* (Ottawa, Jan., 1950).

House of Commons Debates, 1871, 1878, 1879, 1880–1, 1898, 1912, 1939.

House of Commons, session 1937, Special Committee on Farm Implement Prices, *Minutes of Proceedings and Report* (Ottawa: King's Printer, 1937).

House of Commons, Special Committee on Bill 98, Canadian Grain Board Act, *Minutes of Proceedings and Evidence* (Ottawa, 1935).

House of Commons, Special Committee on the Marketing of Wheat and Other Grains under Guarantee of the Dominion Government, *Minutes of Proceedings and Evidence* (Ottawa, 1936).

Report of the Canadian Wheat Board: Crop Years 1935–36 to 1954–55 (Ottawa, annually).

Report of the Canadian Wheat Board, Season 1920 (Ottawa: King's Printer, 1921).

Report of the Commission to Enquire into Trading in Grain Futures (Ottawa: King's Printer, 1931).

Report and Evidence of the Royal Commission on the Shipment and Transportation of Grain, 1900, Canada, *Sessional Papers*, 1900, nos. 81–81a.

Report of the Royal Commission on Dominion-Provincial Relations (Ottawa: King's Printer, 1940).

Report of the Royal Commission on the Grain Trade of Canada, 1906, Canada, *Sessional Papers*, 1908, no. 59.

Report of the Royal Commission to Inquire into Railways and Transportation in Canada [*Drayton-Acworth Report*] (Ottawa: King's Printer, 1917).

Report of the Royal Commission on Price Spreads (Ottawa: King's Printer, 1937).

Report of the Royal Commission on the South Saskatchewan River Project (Ottawa: Queen's Printer, 1952).

Report of the Royal Grain Inquiry Commission (Ottawa: King's Printer, 1925).

Report of the Royal Grain Inquiry Commission, 1938 (Ottawa: King's Printer, 1938).

Statutes.

Alberta

Report of the Commissioner on Banking and Credit with Respect to the Industry of Agriculture in the Province of Alberta, D. A. MacGibbon, Commissioner (mimeo., Edmonton, Nov. 4, 1922).

Sessional Papers, 1927, no. 20 (Edmonton: King's Printer, 1927).

Statutes.

Manitoba

Proceedings of the Conference on Markets for Western Farm Products (Winnipeg, Jan., 1939).

Statutes.

Ontario

RICHARDSON, A. H., *A Report on the Ganaraska Watershed* (Toronto: King's Printer, 1944).

Saskatchewan

Legislature of the Province of Saskatchewan, session 1939, Select Special Committee on Farm Implement Prices and Distribution, *Report* (Regina: King's Printer, 1939).

Report of the Agricultural Credit Commission of the Province of Saskatchewan (Regina: King's Printer, 1913).

Report of the Elevator Commission of the Province of Saskatchewan (Regina: King's Printer, 1910).

Statutes.

STEWART, JAMES, and F. W. RIDDELL, *Report to the Government of Saskatchewan on Wheat Marketing* (Regina: King's Printer, 1921).

United Kingdom

FORRESTER, R. B., *Report upon Large Scale Co-operative Marketing in the United States of America* (London: Ministry of Agriculture and Fisheries, 1925).

ANNUAL REPORTS

Canadian Co-operative Wheat Producers Limited (Winnipeg).

Manitoba Pool Elevators Limited (Winnipeg).

Sanford Evans Statistical Service, *Grain Trade Year Book* (Winnipeg).

Saskatchewan Co-operative Producers Limited (Regina).

Saskatchewan Co-operative Wheat Producers Limited (Regina).

Index

CANADIAN UNIVERSITY PAPERBOOKS
of related interest:

UNIVERSITY OF TORONTO PRESS